Practical Guide for Implementing Secure Intranets and Extranets

For a listing of recent titles in the *Artech House Telecommunications Library,*
turn to the back of this book.

Practical Guide for Implementing Secure Intranets and Extranets

Kaustubh M. Phaltankar

Artech House
Boston • London

Library of Congress Cataloging-in-Publication Data
Phaltankar, Kaustubh M.
    Practical guide for implementing secure intranets and extranets / Kaustubh M.
Phaltankar.
            p. cm. — (Artech House telecommunications library)
    Includes bibliographical references and index.
    ISBN 0-89006-447-4 (alk. paper)
        1. Intranets (Computer networks). 2. Extranets (Computer networks).
    3. Computer networks—Security measures. I. Title. II. Series.

TK5105.875.I6 P43 1999   99-045834
005.8 21—dc21   CIP

British Library Cataloguing in Publication Data
Phaltankar, Kaustubh M.
    Practical guide for implementing secure intranets and
    extranets. — (Artech House telecommunications library)
    1.Wide area networks (Computer networks) - Security measures
    I. Title
    005.8

    ISBN 0-89006-447-4

Cover design by Lynda Fishbourne

International Standard Book Number: 0-89006-447-4
Library of Congress Catalog Card Number: 99-045834

10 9 8 7 6 5 4 3 2 1

To Shadong, my loving wife, and to my parents

Contents

Foreword

Computer-mediated communication has become a business necessity for a significant fraction of businesses in the United States and elsewhere. Internet has been one factor in this growing phenomenon along with some of its critical applications such as electronic mail and World Wide Web services. Electronic commerce is binding companies to buyers and suppliers in an increasingly intimate web of digital relationships. It is not surprising, therefore, that the matter of security has become increasingly important to everyone that depends on computer communication to conduct the course of everyday business.

Security is an overloaded term, meaning many things to many people. For some it represents privacy of communication, for others integrity and authenticity of exchanges, and for others it means freedom from interference and denial of service. In truth, security is all these and more. Security techniques are relevant to many layers of networked information architectures ranging from the protection of dedicated circuits to the protection of content exchanged between cooperating parties.

In this book, Kaustubh Phaltankar explores the nuances and specifics, in detailed, concrete ways, of designing, constructing, and provisioning secure virtual private networks in the context of network substrates shared among multiple parties, including corporate intranets (virtual private networks that use Internet technology) and extranets (just like intranets except that participants are distinct corporate entities, typically).

It is important to appreciate that secure networking is still very much a work in progress. Technically it is possible to provide end-to-end cryptographic security so as to deny an opponent the opportunity to intercept and modify traffic between end parties. However, for this to work, one needs standard encryption algorithms, special cryptographic key distribution mechanisms, and other features related to security. The standards associated with these functions are still not fully agreed by all interested parties. It does seem timely, however, to harvest the large investment made thus far on security techniques in many corporate networks.

Making some of these ideas work in multi-network environments may well require something similar to ideas explored in Public Key Infrastructure concepts developed in the IETF, as well as in various businesses dealing with secure services.

There can be little doubt that demand for this capability is growing quickly, especially as IPv6 makes its debut and forces the focus of collateral attention onto IPSEC—the Internet Engineering Task Force's way of achieving end-to-end packet mode security.

In addition to the essential technology with which this book will make readers acquainted, the case studies are among the most valuable of the contributions the book makes to the current network security literature. Understanding how real companies in the real world have dealt with real security challenges is among the most valuable of the lessons that readers can take away from this useful volume.

Vinton G. Cerf
Camelot 1999

Preface

I hear and I forget,
I see and I remember,
I do and I understand.

Confucius

I have personally followed this mantra all my life. Hence, it is always important for me go beyond the theory and get my hands dirty. I guess that is one reason I became an engineer. I continue to preach this mantra in my professional life and in this book. I wrote this book as a how to guide rather than a technology primer. Over one-third of the book is dedicated to case studies, and the theory section is full of practical ideas and tips based on my personal experience.

I have been working with the Internet before it hit the mainstream around 1995. Since then, the Internet rollercoaster has been nonstop fun and work. What started as fun is now mission critical in many organizations. In the past two years the question has changed from "Is the Internet ready for business?" to "Is business ready for the Internet?" As organizations have realized the power of the Internet and Internet technologies like Internet protocol (IP) and the Web, numerous business applications have mushroomed overnight. Applications for intranets and extranets are taking a center stage in leading the charge for increased information flow, higher productivity, and increased profits. Numerous articles in the press, magazines, and trade journals have confirmed that businesses are ready for the Internet and charging ahead to grab a stake on this new frontier. With the early success of business-to-business e-commerce initiatives, businesses have embraced extranet technology using the Internet as the ubiquitous medium of choice.

In the past two years a number of books have appeared that explain how to create an Internet presence or an intranet business application. These books do an excellent job of introducing the readers to the management and application aspects of building these systems. Once the management has signed off on the concept, there is a significant amount of work that needs to be completed in developing and deploying these applications. The necessary network and security infrastructure needs to be in place before these applications can be deployed in an intranet or an extranet environment. This book addresses that need.

The main focus of this book is to provide practical hands-on material for developing and deploying a high-resiliency network and security infrastructure for mission-critical intranet and extranet applications. The book answers the following questions:

- How to architect a secure, reliable, high-performance, and cost-effective network and the security infrastructure for an intranet and/or an extranet?
- How to configure the various network and security elements, such as routers, switches, servers, and firewalls, to achieve the goals stated above?

The book is full of practical advice based on real-world experience. It follows a logical progression and provides all the building blocks for developing a scaleable and reliable network and security infrastructure. The case study highlights this principle by progressively building the corporate intranet for a fictitious company from one location operation to a global enterprise. We focus on the aspects of legacy network and application integration as well as use of cutting edge virtual private network (VPN) technology for providing remote access to mobile employees, telecommuters, and extranet partners. The case studies focus on the "I do and I understand" principle by providing complete configuration details of routers, firewalls, servers, and VPN devices. Each case study builds on the implementation of previous case studies. Each implementation uses the most prevalent technologies in the areas of local area network (LAN), wide area network (WAN), routers, firewalls, and VPN. The implementations are built using products from leading vendors like Cisco, Checkpoint, Motorola, and Aventail.

Intended Audience

The book is intended for a broad range of audiences interested in the implementation aspects of intranets and extranets, including:

- IS and IT professionals who have been tasked to implement the network and security infrastructure;
- Managers who would like to understand the practical aspects of selecting the right technology and vendors for the project;
- Designers and architects who have been tasked with providing the technology and business vision for the Internet, intranet, and extranet strategy;
- Students and professionals who plan to play a role in the Internet industry and would like to understand the behind-the-scene aspects of implementing secure intranets and VPN tunnels.

Book Layout

The book is organized to provide the technology aspects first, followed by the case studies. Each chapter can be read separately or sequentially.

- *Chapter 1, Introduction:* Introduces the concept of Internet, intranet, and extranet. The chapter explores the traditional method of connecting intranet sites and the new approaches based on frame relay and Internet VPN. We close the chapter with an introduction to extranet, its advantages, and security requirements.
- *Chapter 2, Wide Area Network Components:* Discusses design options for building resilient WAN connections in deployment of regional or global intranets and extranets. The chapter covers the WAN technologies and topologies.
- *Chapter 3, Local Area Network Components:* Discusses the design options for building a resilient LAN infrastructure. LAN technologies like ethernet, fast ethernet, gigabit ethernet, ATM, FDDI, token ring, and LAN switching are explored with the LAN routing protocols. The goal is to provide various LAN design options and their selection criteria.
- *Chapter 4, Network and Service Management:* Describes the technologies available for network and service management of an intranet or an extranet service, including discussion of SNMP and RMON MIBs to monitor internal and external service level agreements (SLAs).

- *Chapter 5, Security Components of Intranets and Extranets:* Introduces the crucial elements of secure intranet and extranet architecture and deployment. We begin the discussion on the security framework and then explore technologies available for providing data security, access control, and authentication. The concept of digital certificates is introduced next with discussion on the use of public key infrastructure (PKI), global directory, and global single sign-on solutions followed by a discussion on different types of firewalls and use of network address translation (NAT).

- *Chapter 6, Virtual Private Network:* This chapter provides an extensive discussion on the what, why, and how of VPN. This discussion is followed by a unique exploration of VPN technologies currently available in the market at various layers of the OSI stack. This includes IPSec, GRE tunnels, PPTP, and L2TP tunneling protocols for LAN-to-LAN and dial-up VPN connections over the public Internet.

- *Chapter 7, Case Studies:* The case studies provide the real-world implementations of intranet and extranet network and security infrastructure. The case studies provide installation and configuration details for routers, switches, remote access devices, Internet VPNs with IPSec, VPN clients, VPN servers, frame relay PVCs, and SNA gateways in a legacy environment.

There is a tremendous amount of practical information that can be readily applied to your environment. Though I have tried to cover the most typical scenarios, your specific situation may be different. The concepts, theories, examples, and references in the book should provide enough material to formulate a solution that meets your custom requirements.

Web Site

Additional data has been provided on the Web site http://www.netplexus.com/ artechhouse. The Web site allows us to maintain a dynamic repository of information on each topic discussed in this book. Internet and Internet technologies are evolving every day and providing added impetus to the growth of organizational intranets and extranets. The Web site will allow you to continuously tap into the latest information on technology and products that could be helpful for your intranet and extranet implementations. I hope that you will find this site useful. Any comments or feedback are welcome.

Acknowledgments

I must thank all my team at MCI and later at Cable & Wireless for providing a continuous learning environment that encourages individuals to get their hands dirty. Special thanks to my mentors Dick Stephens and Randy Catoe for their continuous support during my days at MCI. Jim Martin and the operations team at the hosting center have provided continuous customer feedback to get the technology working for the customers. Dale Drew and Steve Weeber from security engineering provided early support in my forays into the world of security while Dr. Abdou Youssef provided a detailed feedback on case studies. My special thanks to Dr. Vint Cerf and John Clenson at MCI, for keeping me on my toes. Every conversation with them makes me realize how much I still have to learn.

I would also like to thank the acquisition editors, Mark Walsh and Barbara Lovenvirth, at Artech House, for being patient with me during the long process of manuscript acquisition. Last, but not the least, I would like to thank my wife, Shadong, for constantly encouraging me to complete the project, even though we were expecting our first born. Without her help and constant motivation this project may have remained in my dreams only!

Conventions

a. Symbols used

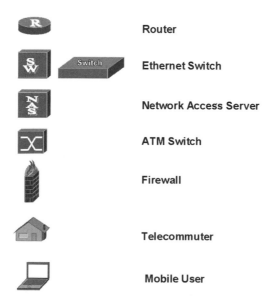

Router

Ethernet Switch

Network Access Server

ATM Switch

Firewall

Telecommuter

Mobile User

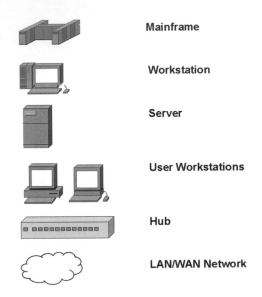

Mainframe

Workstation

Server

User Workstations

Hub

LAN/WAN Network

b.   Type conventions

no ip   Indicates Router Commands

The interface   Indicates Explanatory Comments

Kaustubh Phaltankar
E-mail: kaus@netplexus.com
Web: www.netplexus.com

1

Introduction

In this chapter, we will define the terms *Internet, intranet, and extranet,* followed by examples of each category. At first we will briefly draw a distinction between Internet and Internet-related technologies and then explore the technologies and concepts related to intranet and extranet in greater detail. Next, we will focus on the advantages of intranets and extranets, followed by their network and security requirements including various implementation options.

Definitions:

- *Internet:* The Internet is a global network of interconnected networks, connecting private, public, and university networks in one cohesive unit.

- *Intranet:* An intranet is a private enterprise network that uses Internet and Web technologies for information gathering and distribution within an organization.

- *Extranet:* An extranet is a community of interest created by extending an intranet to selected entities external to an organization.

1.1 Internet

In this book, our assumption is that you are already familiar with the Internet and have access to numerous articles and books that explain its origin and evolution. Based on this assumption, we will briefly introduce the Internet and then focus on the areas of intranet and extranet.

As defined above, the Internet is a global network of interconnected networks as shown in Figure 1.1.

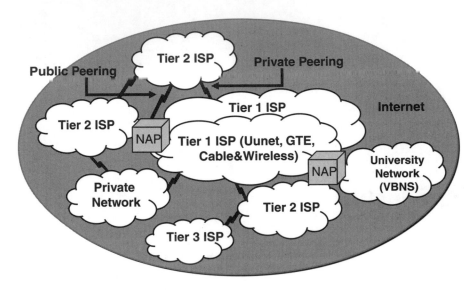

Figure 1.1 Global Internet network hierarchy.

The Internet exists due to the interconnection of private, public, and university networks. The networks connect to each other at various public or private connection points, also known as peering points. The public peering point is known as the network access point, or NAP. The Internet service providers are classified into three tiers based on the size and the capacity of their networks. Tier 1 Internet service providers (ISPs) like UUNet, GTE, Sprint, and Cable & Wireless have large Internet backbones of multiple OC-12 (655 Mbps) or OC-48 (2.4 Gbps) capacity. Most of the tier 1 service providers maintain connections to the NAPs, thus allowing smaller ISPs to connect to the big ISP networks at public peering points. The public peering point is the most cost-effective method of communicating with the other networks on the Internet. Due to the wide popularity of the public peering points, they have started to become congested. One way to circumvent this congestion problem is to set up private peering arrangements between selective service providers. For example, if traffic between a tier 1 and tier 2 service provider meets certain acceptable use policies (AUP) of the tier 1 service provider, the tier 1 provider will have a private peering with the tier 2 provider. The private peering could be set up at an NAP or at tier 1's point-of-presence (POP).

Note It is a good practice for an enterprise to have multiple connections to the same or different ISPs. The best resiliency is provided with dedicated Internet connections to different ISPs. One can potentially load balance the traffic across multiple connections, thus utilizing both circuits. If you do not have sufficient traffic to justify a second dedicated connection, make provisions for an integrated services digital network (ISDN) or public switched telephone network (PSTN) dial-back circuit that can become activated if the primary circuit reaches a certain traffic threshold or gets disconnected. The traffic can then be diverted to the backup connection in either scenario.

Before we discuss the definition of an intranet, let's draw an important distinction between the Internet and Internet-related technologies. The term *Internet* relates to the concept of global network as explained earlier, while the term *Internet technologies* refers to a collection of Internet protocols and applications. The Internet is based on TCP/IP (Transmission Control Protocol/Internet Protocol) for network transport and a collection of TCP/IP-based applications like SMTP (simple mail transport protocol) and POP3 (post office protocol version 3), Telnet (for remote login), and FTP (file transfer protocol, for exchanging data files). This collection of protocols and applications is collectively known as Internet technologies.

Another term used in conjunction with Internet technology is *Web technology*. The term *Web technology* refers to the use of HTTP (hypertext transport protocol) for transport of Web data and HTML/XML (hypertext markup language/extended markup language) for data presentation. The HTTP protocol is delivered over a TCP/IP session. A Web server and a Web client (for example, the Netscape browser) communicate with each other using HTTP protocol. The information on the Web server is presented on the Web client using HTML/XML tags. Due to the great popularity of the Web interface and the low cost of Web browser deployment, Web browsers are now a universally accepted client desktop interface.

Next, we will take a closer look at the definition of *intranet* and typical technologies used in building an intranet solution.

1.2 Intranet

As the technology leaders and early adopters have proved the usefulness of an enterprise intranet, the pace of intranet adoption has accelerated. As author Geoffrey Moore put it, we are currently crossing the "chasm" in the intranet adoption curve, separating the "Early Market" from the "Main Stream Market." The main stream market growth in years 2000 and 2001 will bring tornado-like growth to this market. Initially, the early adopters of intranet used the newfound technology to:

- Provide easy access to internal data by publishing the information on departmental intranets;

- Set up employee self-service Web sites for human resources, payroll, sales, marketing, and training.

As the understanding of the new technology has progressed, the adopters have used the technology for more complex applications like:

- Collaborative workflow managers, scheduling, messaging, and discussion groups;
- Inventory and logistic management systems;
- Customer help desk and knowledge management systems.

The successful intranets have allowed organizations to:

- Save money;
- Increase employee productivity and employee retention;
- Empower the customer help desk personnel with intranet knowledge management systems, resulting in increased customer satisfaction and retention;
- Provide competitive advantage by making the product information, production schedules, and the product competitive analysis only a click away. This empowers the sales, marketing, and support group, resulting in increased sales, added revenue, and customer retention.

In most organizations, intranets have started as small departmental efforts. These efforts soon spread like wildfires throughout the enterprise, resulting in pockets of information that may not be accessible to the rest of the organization. Organizations have realized that this type of uncontrolled growth is counterproductive and have started to implement enterprise-wide strategies for an intranet with decentralized content management. This approach has resulted in uniform information delivery mechanisms, while still maintaining a grassroot content development strategy, thus enabling organizations to fully utilize the power of intranets.

An enterprise intranet may be based on Internet and Web technologies in combination with vendor proprietary technologies. Many organizations today have fully functional internal networks based on a legacy network protocol or a vendor proprietary network protocol like IBM's Source Route Bridging (SRB) or Novell's IPX protocol. Though these are functional internal networks, they are not able to utilize the full potential of all organizational resources due to the high cost of deployment, nonintuitive user interface, limited expansion capability, and the proprietary nature of the applications and protocols. These limiting factors can now be addressed with standard protocols like TCP/IP, HTTP, and HTML, and a legacy gateway. Proponents of proprietary technology, especially the vendors, have learned their lessons the hard way. Market shares of IBM and one-time networking giant, Novell, have fallen behind vendors like Sun and Microsoft, who embraced the open standards. Now the same vendors who profited from the proprietary nature of their architecture have started to implement translation protocols and gateways to the Internet standards. This illustrates the phrase, "If you can't beat them, join them." These vendors hope to stem the slide in their market share and capitalize on the popularity of the Internet and Internet technologies. For example, Novell has introduced total IP support in Novell 5.0 while IBM has introduced TCP/IP support for mainframes. Vendors like Open Connect have introduced IP to SNA gateways to provide Web access to the data locked in the proprietary technology of the past.

What are the prerequisites for a successful deployment of an intranet?

- Unified intranet strategy and architecture;
- Secure and ubiquitous intra-enterprise connectivity;
- Integration with legacy networks and applications (including mainframe);
- Use of standards-based technology for network, security, and Web authoring;
- Common user interface based on a universal Web client, which helps in reducing the cost of maintaining each desktop;
- Implementation of enterprise-wide policies and guidelines on network access, security, content authoring, and management;
- Management and user buy-in.

Considering these design requirements, we will now explore the design options for building intranet networks and their security implications. We will first consider the traditional private line option followed by the frame relay option and then conclude with the Internet-based VPN approach for building secure intranet networks.

1.2.1 Traditional Approach

One of the important requirements listed for a successful deployment of an enterprise intranet is the "secure and ubiquitous intra-enterprise connectivity." This implies connectivity for all offices (branch, regional), remote desktops (telecommuters), and mobile users. This is a tall order. Traditionally, organizations have adopted an architecture that is similar to the one shown in Figure 1.2.

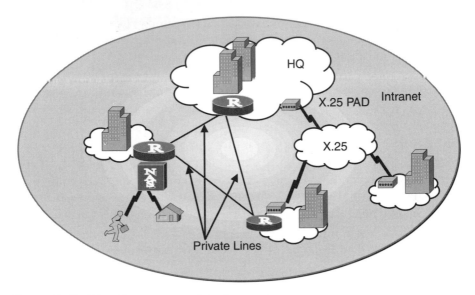

Figure 1.2 Traditional intranet network implementation.

Each regional office would be connected to the corporate headquarters using private-line connections. Remote branch offices and international offices are connected through a low-speed X.25 or dial-up connections. Many multi-national corporations like Hewlett Packard and Citicorp have built their own private networks spanning the globe. The pros and cons of the private line approach are given below.

Pros:

- The networks are completely controlled by one entity.
- The networks are deemed to be secure due to this tight control by one entity. This control allows the company to set its own policy and the associated security measures.
- The direct point-to-point connections using the digital private lines provide an error-free transmission medium for higher speeds.

Cons:

- The cost of a private line is distance sensitive; hence, it becomes very expensive to build-out a widely distributed network.
- The large cost of deployment also prohibits any-to-any connectivity for all offices.

- The cost of maintaining a large technical team to handle installation and troubleshooting is very high.
- Even if the network is controlled by one entity, over a period of time the security implementation develops major security exposures. For example, exposures are due to changing personnel, lost or forgotten passwords, and lack of a strong authentication scheme.

As for the X.25 connections, they have traditionally been low-speed connections (up to 64 Kbps). Hence, these connections cannot handle the ever-increasing need of bandwidth-hungry applications. However, this scenario is now changing due to improvements in technology (up to 2 Mbps).

Each location maintains its own dial-modem bank for providing remote access to its mobile users and telecommuters.

1.2.2   Frame Relay–Based Approach

During the late 1980s and early 1990s, the end devices, PCs, and workstations started to become more intelligent and the need for LAN-to-LAN networking was beginning to be felt. With the need for ubiquitous low-cost connectivity to meet the needs of bandwidth-hungry LAN-to-LAN applications, corporations started to turn to frame relay networks for their WAN connections. Frame relay originated in narrow-band ISDN, but later became a viable alternative to private-line connections. Unlike the private-line technology, frame relay is a shared service. The shared-service infrastructure is provided by existing telephone companies like MCI Worldcom, AT&T, and British Telecom. The point-to-point connectivity is based on frame relay–based permanent virtual circuits (PVCs). Figure 1.3 shows an intranet implementation connecting the various regional and branch offices of an enterprise, using a frame relay network.

Note   We are, of course, simplifying the history of the transition to frame relay networks. For more details on the history of frame relay, please refer to the Bibliography.

Frame relay customers buy ports of certain capacity for each location from the service provider. The customer premise equipment (CPE), like routers connecting to the frame cloud, are configured with PVCs. Each PVC is then configured with a local address, also known as data link connection identifier (DLCI), and other bandwidth parameters like the committed information rate (CIR) and committed burst rate. Figure 1.3 shows a high-resiliency network with two PVCs connecting the non-HQ (headquarters) locations to the HQ. The HQ has two separate physical connections to the frame relay cloud for redundancy. The non-HQ location can either load share the traffic across the two

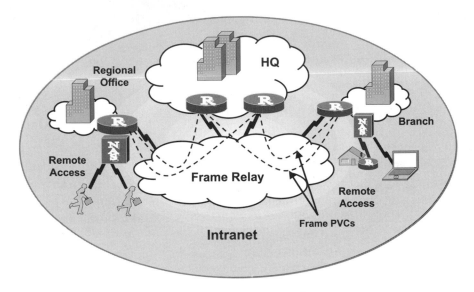

Figure 1.3 Intranet network implementation using frame relay.

PVCs or use them in redundant mode. Depending on the traffic between the regional office and the branch office, a third PVC directly connecting the two locations could be added to the network configuration, thus avoiding the hop through HQ. As in the previous configuration, each location maintains its own dial-modem bank for providing remote access to its mobile users and telecommuters. What are the pros and cons of the frame relay–based intranet approach?

Pros:

- Efficient for LAN-to-LAN data networking.
- Universal LAN-to-LAN network connectivity at a lower price than the private-line option.
- Higher reliability at a modest cost. The frame relay cloud is engineered by the service providers to be highly resilient. The cloud engineering provides for rerouting capability in case of a failure within the cloud. As for a site resiliency, the incremental cost of adding a backup PVC is quite minimal.
- Frame relay provides same or better performance as compared to the private-line option at 30%–40% cost savings.
- It provides a high level of security, since the traffic is limited to the PVC.

- It can support multiprotocol transport based on RFC (request-for-comment) 1490.

Cons:

- It has variable delays, not suitable for interactive multimedia applications with audio and video content.
- Frame relay connections are not scalable. A full mesh requires N (N-1)/2 circuits.
- Dynamic connections to new sites cannot be established on-demand. Adding a new site to a mesh requires installation of a new circuit and new PVCs. New circuit installs have a long lead-time depending on the availability of port capacity on the provider network and availability of local loops from the local exchange carrier (LEC). New frame relay services based on switched virtual circuit (SVC) have made some headway in this direction.
- During congestion, frame relay-based traffic cannot be configured to discriminate between critical and noncritical traffic.

1.2.3   Internet VPN–Based Approach

How do you provide dynamic any-to-any connectivity? How do you provide a highly reliable and scalable network connection that can handle the requirements of high bandwidth, interactive multimedia traffic? The next evolution in the mid-1990s was the emergence of Internet as an answer to the limitations of frame relay networks. The emergence of new markets for Internet-based applications like Web, news, mail, and FTP provided added impetus to the growth of the Internet market. New areas of electronic commerce continue to drive the hypergrowth of the Internet market. Public network carriers and telephone companies have spent millions of dollars in building the Internet infrastructure. We have already discussed the Internet and Internet-related technologies. The question we want to explore now is, "How can we leverage this huge Internet infrastructure for building a secure, universally accessible, network infrastructure for private enterprises?" One way to build privacy in the connection is to encrypt the communication while it travels over the public Internet. Further, you can authenticate the end points and check for data integrity when the data is received by the receiver, thus benefiting from the universal access provided by the public Internet without sacrificing the security. One of the techniques used to implement such secure connections over the public Internet shares quite a few characteristics with a frame PVC. This technique is known as virtual private network (VPN). An example of an intranet created by configuring VPN connections over the public Internet is shown in Figure 1.4.

In Figure 1.4 the Internet cloud could be associated with one or many ISPs. Each site is then connected to the ISP network with a dedicated point-to-point connection. The ISP could be national or regional. The firewall at each location protects the individual site and enables organizations to manage the interface between the private and public part of the network. Some firewalls also enable setup of virtual private networks over the Internet. The end-to-end VPN connections are shown by dotted lines. Each location can use its VPN connection to connect to remote sites or connect to other office locations over the Internet, as shown in Figure 1.4. We will look at different types of firewalls and VPN components in Chapter 6.

The remote LAN access servers have been positioned closer to the user in the network, thus saving organizations the cost of 800 long distance calls. The network access servers (NAS) could be owned by organizations or by ISPs.

The Internet-based approach has its own pros and cons:

Pros:

- Interoperability with other Internet users and Internet and Web technology-based applications due to standards-based technology.

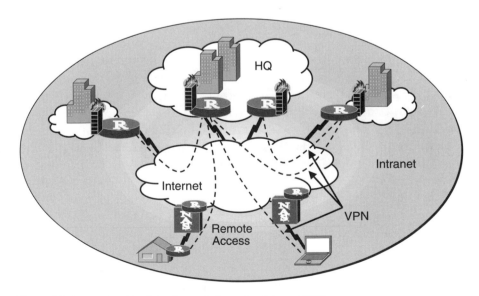

Figure 1.4 Intranet network implementation using Internet VPNs.

- Universal low-cost connectivity for LAN-to-LAN communications.
- Reduced cost of supporting mobile and telecommuting users.
- Provides high-speed data connections for interactive multimedia applications.
- Application level discrimination of traffic at the edge allows administrators to configure various levels of quality of service (QOS) parameters.

Cons:

- Because of the public nature of the Internet, security is always a threat.
- Since a data packet might traverse multiple service provider networks, there are no guarantees on performance. The data is handled on a best-effort basis. This implies that organizations have to maintain separate links for time-sensitive mission-critical data like SNA communication to mainframe.
- Added complexity of firewall and VPN management.

As Internet service providers continue to make large investments in the Internet backbone and new technologies like multipath label switching (MPLS), they will be able to offer better service level agreements (SLAs) in the future. UUNet and GTE are currently offering 99.95% availability of backbone and 130–150 ms of network latency between any two nodes on their backbone. In the meantime, it is recommended that organizations have a mixed strategy where the mission-critical data flows over a private network like a frame relay network, while all other data is transitioned over the Internet using virtual private network backbone.

1.3 Intranet Components

We will use the typical intranet setup shown in Figure 1.5.
We can clearly identify five major components of an intranet setup:

1. The WAN component;
2. The LAN component;
3. The server and application software component;
4. The security component;
5. The intranet service management component.

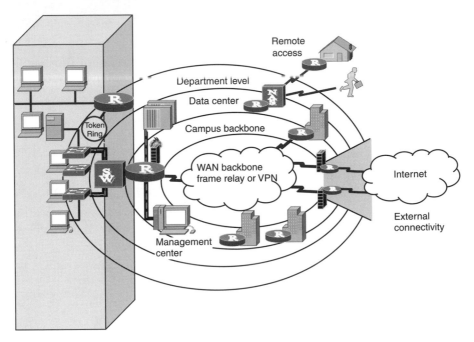

Figure 1.5 A sample intranet architecture.

The WAN Component

The WAN component consists of:

- The physical connection(s) to the Internet;
- The physical connection(s) to a frame relay cloud;
- The WAN router and network addressing scheme;
- The configuration of frame relay PVC on a WAN router;
- The configuration of Internet-based VPN on a WAN router or WAN firewall;
- The WAN network routing protocol like BGP for external connection and OSPF for internal connection.

The LAN Component

The LAN component consists of:

- The LAN technology;
- The LAN topology;
- The LAN router and network addressing scheme;
- The LAN network routing protocol;
- The LAN firewall and load-sharing components.

The Server and Application Software Component

The server and application software component consists of:

- The intranet application servers and associated clients;
- The gateway server for protocol translation;
- The high-resiliency design of the server component for high availability;
- The backup server and associated backup hardware like the tape jukebox.

The Security Component

Literally translated, *intranet* means "intra = internal," "net = network." An intranet is a private enterprise (internal) network that is kept private by providing strict access control from external networks like the Internet. The access control is achieved by using security technologies like "air gap" or a "firewall." Please refer to Figure 1.6.

In the case of an "air gap," the internal network is not connected to the external network at all. This security by isolation has been a traditional means of maintaining security in many organizations and continues to be practiced by quite a few organizations in medical and financial communities, even today. In some cases organizations have implemented two completely separate networks for internal and external communications. Most other organizations, however, see some need for connecting their internal networks to external networks like the public Internet or a vendor/partner network. In such cases, firewall techniques are used to protect the security and privacy of their internal network from the external entities.

Note Various firewall techniques are explained in detail in Chapter 5.

The security component consists of:

- The corporate security framework and the security policies for internal and external connectivity;
- The LAN and WAN firewalls and VPN setup;

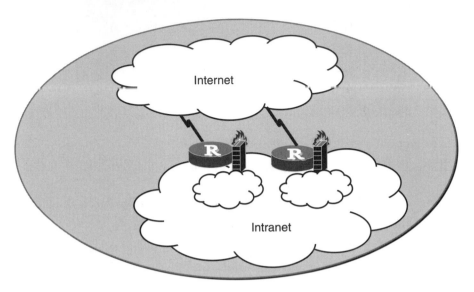

Figure 1.6 Access control using a firewall.

- The secure configuration of the intranet application and the server operating system;
- The encryption, authentication, and data integrity check methodologies.

The Intranet Service Management Component

The intranet service management component consists of:

- The management of intracompany SLAs;
- The management of SLAs with external service providers like ISPs and
  frame relay service providers;
- The internal network management and reporting systems;
- The internal help desk and trouble ticketing system.

We will cover each one of these components in greater detail in the next
four chapters. Our order of coverage is shown in Figure 1.7.

1.4 Intranet Summary

With the rise in high-bandwidth interactive multimedia applications, the demand for high-bandwidth LAN-to-LAN connectivity has grown exponentially.

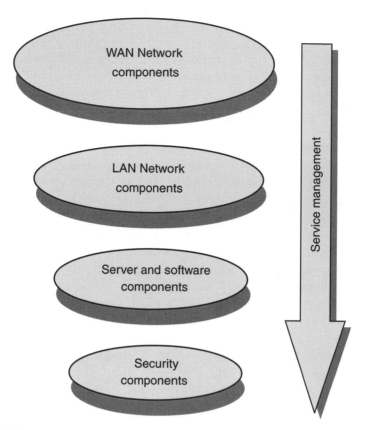

Figure 1.7 Intranet components.

Network designers have always looked for solutions to support the ever-growing bandwidth demand at the lowest possible cost. With the advent of Internet VPN and various packet-switched technologies like frame relay and asynchronous transfer mode (ATM), network designers can now build on the vision of ubiquitous secure access in the most cost-effective manner.

Even though, both the Internet and intranets are based on the Internet and Web technologies, when it comes to implementation, user expectations differ significantly. Intranet implementations have more stringent SLAs on application response time, network latency, data, and access security, while the Internet access is still on a best-effort basis. The requirements on SLAs become more stringent when we implement extranets. Let's explore the extranet concept in further detail.

1.5 Extranet

Extranet is a community of interest created by "extending an intranet" to selected entities external to an organization.

In the definition, we are very careful to keep the relationship between various entities at a more general level, since a relationship could exist between any two organizations, private, public, or combination thereof. Though the definition is generic, almost all early implementations of extranets have been in the private sector. These private sector extranets have been designed to foster corporate business-to-business relationships between vendors, partners, and customers. Most of these extranets have been set up over public network infrastructure using Internet technologies like TCP/IP and the Web. Given the choice of ATM, frame relay, and Internet-based switched-network infrastructure, the Internet-based extranet implementations have been most popular due to their low cost of implementation, ubiquitous access, and improved security offerings based on global standards. In a recent study from Forrester Research, 76% of the companies use the public Internet for their extranet setup and an even larger 88% of the companies plan to do so in next five years. Another set of questions regarding the current deployment of extranet applications showed that a large percentage of these extranets have been initially launched to take advantage of e-commerce opportunities (see Series S1 in Figure 1.8).

In Figure 1.8, the first series (S1) represents answers to the survey question, "What application drove the launch of your extranet?" while the second

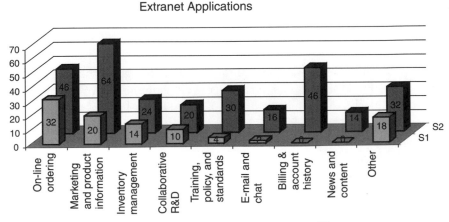

S1: What application drove the launch of your extranet? ■
S2: What applications are you running now? ■

Figure 1.8 Extranet driver applications. (*Source:* Forrester Research, Inc.)

series (S2) represents answers to the survey question, "What applications are you running now?" What is interesting to see is that what started as an on-line e-commerce initiative has quickly diversified into back-office applications like billing and accounting, inventory management, and marketing and product information.

Other applications of extranet include:

- Access to legacy database and mainframe;
- Enterprise resource planning (ERP) and supply chain management;
- Collaborative applications like schedule management, interactive conferencing, and discussion groups;
- Customer service and product support;
- Customer "self-service" applications;
- Bulletin boards and closed user groups (CUG);
- On-line financial transactions.

1.5.1 Advantages of Extranet

Extranet brings many advantages and benefits to these mission-critical applications. For example:

- Collaborative R&D fosters better ties between partners and shorter time-to-market thus reducing the product life cycle and increasing the competitiveness of products.
- ERP and supply chain management can help in streamlining the business processes in manufacturing, customer service, and product development resulting in better business efficiencies, reduced inventories, reduced cost of production, and increased competitiveness. The bottom-line: a better product at a lower cost, resulting in higher market penetration and larger profit margins.
- Better customer support results in increased customer loyalty, while the customer "self-service" applications provide quality 7×24 support throughout the year at a lower cost.
- Collaborative extranet applications help to foster better team spirit among partners and customers, improving the speed of communication. Collaborative applications also reduce the cost of production, reduce travel costs, and increase the efficiency of business development process.

- Customer and partner access to internal company data and their participation in internal business processes helps in building loyalty, streamlining development cycles, and creating barriers for competitors.
- Participation in extranets also brings access to new markets. If a company could not afford the high cost of electronic data interchange (EDI) transaction in the past, new Internet-based extranet EDIs can make company participation affordable and bring new markets and customers where the company could not compete before.

With all these wonderful advantages, where does one start and what should one consider while taking the first steps toward building extranets?

One of the first things to recognize is that extranets provide a "controlled" access to internal resources from external entities. Thus, in most cases, access to extranets is by invitation only. Except for a few exceptional categories, like the FedEx extranet or an airline reservation system, you need to prenegotiate terms of extranet access privilege with your vendors, customers, partners, suppliers, distributors, and contractors. The negotiation phase assumes that your partners and customers have bought off on the extranet concept. This early buy-off is very important since extranet implementation will involve a change to the current familiar business processes within your organization as well as your extranet partner's organization. Even though the current processes may be manual and slow and use legacy systems, they are preferred due to their familiarity. Since any change in the known system involves retraining people, new ways of thinking, and new processes for getting things done, there may be opposition to these changes. Hence, it is important to convey your plans for implementing extranets to all interested parties as early as possible and to get their continued involvement throughout the design and implementation phase. Some of the important considerations while developing extranet network, security, and application designs are:

- Security exposure to internal and partner networks and systems;
- Ease of use;
- Interoperable technology;
- Scalability;
- Ability to leverage legacy applications;
- High availability (as the extranet applications become mission critical, it is important to maintain the constant availability of these applications);

- Performance of the network and the extranet application (if the extranet adds more cycles to a product development process than reducing them, it is not serving its purpose and will be opposed);

- Setting correct expectations (for example, customer services. If the customers perceive reduced support rather than increased support, due to lack of human interaction or ill-designed application, it is going to turn off customers instead of retaining them);

- Legal considerations of connecting multiple systems and the potential impact on the current operation due to this new setup should be explained and agreed upon in advance.

Let's take a detailed look at some of these important considerations.

1.5.2 Security

The most important aspect that should top everyone's list is security. When two or more organizations collaborate on an extranet, the security concern extends both ways (see Figure 1.9).

In the extranet setup shown in Figure 1.9, the extranet-sponsoring organization on the left is opening the doors to its intranet applications and needs to provide tight security with strong authentication to prevent any unauthorized use of its resources. In this scenario, since the access is one-way, the security design at the supplier or partner network is different and slightly less complicated than the one at the gates of the sponsoring organization. This model also could be generalized to the Internet-based extranet model of FedEx, where external users on the Internet can access the package tracking information on the systems on the FedEx intranet. An alternative extranet model is where there is collaborative extranet setup among all the parties with two-way communication as shown in Figure 1.10.

In this setup, all three parties participating in the extranet have an equal risk factor since they allow two-way peer-to-peer communication between each participant's intranet. The type of extranet setup will, of course, be dictated by the relationship between each party and the goals of the extranet connectivity. In some extranets, access to vendors and partners may be restricted to an application server with replicated data (see Figure 1.11) in the demilitarized zone (DMZ). (DMZ is a boundary network separating internal and external networks.)

The extranet application server in the DMZ receives the extranet application requests from external users and either processes the data locally based on the replicated content from intranet, or retrieves the data dynamically from the intranet servers. The access to the extranet application server can be restricted to the partners and suppliers by:

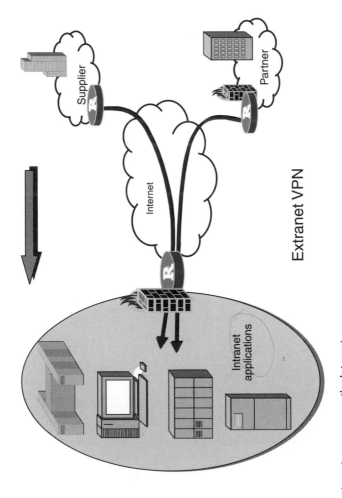

Figure 1.9 Inbound extranet access over the Internet.

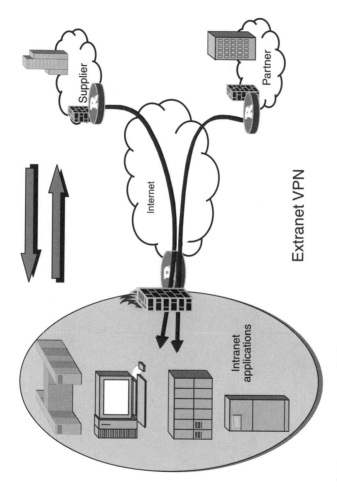

Figure 1.10 Bidirectional extranet access over the Internet.

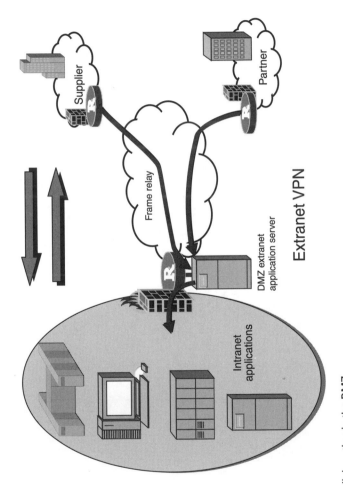

Figure 1.11 Extranet collaboration in the DMZ.

- Using access control lists on the router.
- Using an authenticated application proxy via the firewall server shown in Figure 1.11. In this scenario, the application server would be connected to one of the interfaces on the firewall server, instead of an interface on the router as shown in Figure 1.11. The authenticated proxy server solution could be based on a Microsoft proxy server, Checkpoint Firewall-1 proxy solution, or Aventail's SOCKS5-based proxy solution.

The choice of the solution depends on the desired level of access control and audit tracking capability. If you add to this solution a VPN setup over the public Internet, you will also need to consider issues like encryption, key exchange, and extranet performance. We will explore each one of these aspects in detail during the discussions on security, VPN, LAN/WAN network design, and the case studies with real implementations. In the end, we will wrap up with actual implementations of intranet and extranet design including the router, switch, and firewall configurations.

Other extranet considerations like ease-of-use, use of interoperable technology, scalability, ability to leverage legacy applications, and high availability are equally important. A well-designed extranet application and a network/security design will help you in achieving the desired return-on-investment (ROI) on the extranet project. In most cases, the payback time is expected to be in the range of one to four years.

Note on ROI   In the case of the extranet it is quite a challenge to quantify the return-on-investment. The gains due to an extranet implementation are spread across multiple internal and external boundaries. A satisfied customer can lead to increased customer loyalty, lower turn around, and reduced cost of retaining and acquiring new customers, which in turn can lead to reduced cost of doing business resulting in higher profits, thus greater ROI. As you can imagine, it is a tough task to quantify the gains in each area. Until such time that mechanisms can be put in place to capture the savings in each phase, you should look at the overall picture of a satisfied customer or a happy supplier in deciding on the extranet deployment plan. An early buy-off from the upper management, vendors, and partners is a must!

1.5.3   Examples of Extranet

In the definition of extranet, we used the concept of "community of interest." Now we will take a look at some of the examples of such extranet communities of interests.

Once such community of interest could be a collection of vendors and partners. Figure 1.12 shows two such examples of extranet communities.

In one case the extranet community consists of a vendor, partner, and the XYZ Corporation, while in other case it includes the XYZ Corporation and its customer. In the first case, the three members of community can now collaborate and exchange information within the secure boundaries of the extranet. An example of such collaboration would be that a vendor of XYZ Corporation provides support information directly to a partner of XYZ Corporation, while the partner collaborates with the XYZ Corporation regarding the rollout of a new product. To accomplish the goals of such an extranet, the XYZ Corporation can set up a collaborative working environment by deploying collaborative extranet applications with shared and routable information repositories that can be accessed by all interested parties. The XYZ Corporation can also provide limited access to corporate intranet applications and databases. The access would be limited to the members of the extranet community and can be tailored to meet the level of authorization and the scope of access granted to a member of this community. The corporate security policy on the extranet will have to define the security framework, the corporate policy, and the audit requirements.

At the network level, the extranet connectivity can be accomplished by way of a VPN over the Internet or a PVC over a public frame relay network. Both types of setup are illustrated in Figure 1.12. One of the best examples of

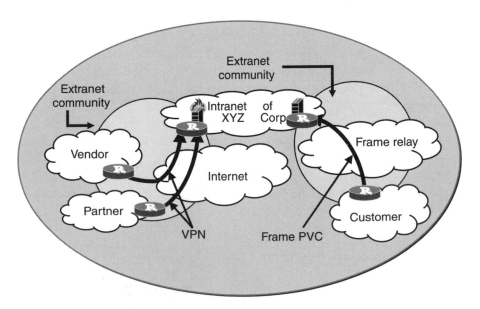

Figure 1.12 Extranet community of interest (COI).

present day extranet is the automotive network exchange (ANX) extranet. The ANX network has been sponsored by the big three auto (now big two) manufacturers and their 13 major suppliers. The main intent of the ANX extranet is to enable better information flow within the manufacturing supply chain process. The new extranet process flow is expected to save almost $1 billion in the first year alone. The time spent on the supply-chain processes has also been reduced from four weeks to four days, resulting in major productivity improvements and reduction in turnaround time for getting a car rolled out of the assembly line. Figure 1.13 shows the ANX network setup.

The ANX network is an isolated network that has been carved out of each CSP's network. The current CSPs include Bell Canada and Ameritech. The ANX CEPO is isolated from the Internet. Any participating CSP has to adhere to very strict security, routing, and performance requirements. The ANX CSPs act as the network entry points for the ANX trading partners. The ANX trading partners could buy a direct port on the CSP network or connect to the CSP over the public Internet using IPSec-based VPN. This extranet allows the trading partners to participate in the complex supply chain processes of the auto manufacturers.

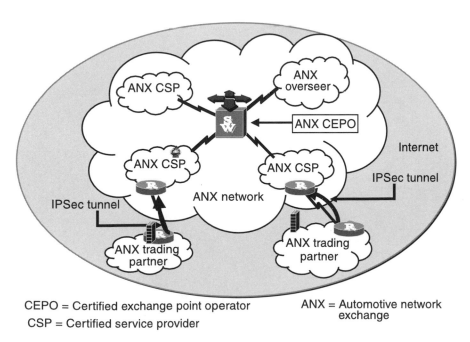

CEPO = Certified exchange point operator
CSP = Certified service provider
ANX = Automotive network exchange

Figure 1.13 ANX extranet.

A typical "supply-chain process" in an organization can be depicted by Figure 1.14.

Extranet-based supply chain has several advantages:

- Since the supply-chain process is almost completely automated several departments get automatic notifications as a user request is processed through the chain. For example, when the item is shipped from the supplier the production manager is notified of the arriving items, while the purchasing manager gets the shipping details and the invoice. When the item is received the inventory is automatically updated to reflect the new quantities of items.

- The extranet supply-chain process allows manufacturers to control inventory costs, while the procurement can do a better job of forecasting and ordering, thus gaining economies of scale.

- The extranet supply-chain implementation is equally applicable to small or big purchases. For small purchases the approval routing process may be different and quicker, thus the time and frustration of going through a lengthy purchasing process for small items is eliminated. This item alone can have a significant impact on corporate savings. When users get frustrated by a lengthy requisition and purchase process, they are likely to bypass the process. Since the items have been

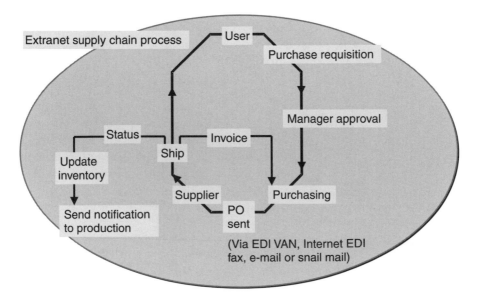

Figure 1.14 Use of an extranet in supply-chain management.

bought outside the normal process, it imposes costly processing fees, resulting in less control over the supply chain and forfeiture of cost savings due to economies of scale.

Software from ERP vendors like SAP, Oracle, and Baan provides an out-of-the-box implementation of extranet applications. Almost all the solutions are now available as Web-based solutions and can be accessed by users using a standard Web browser. For added security, the client-server communication could be encrypted using SSL encryption and authenticated using digital certificates. We will take a closer look at these technologies in Chapter 5 on security. These types of Internet technology–based extranet and intranet process implementations have provided great boost to the on-line business-to-business e-commerce market.

Other successful examples of extranet implementations are:

- McDonnell Douglas's inventory management extranet provides access to 500 suppliers and tracks 80,000 parts and assemblies.
- Shell Chemical Company's extranet takes a unique approach by managing inventories for its customers, improving the operational efficiency, and increasing customer satisfaction.
- Cisco Systems' extranet allows customers to track and order equipment on-line based on a prenegotiated relationship.
- DaimlerChrysler's extranet in Japan allows its dealers to process orders over the extranet.
- Courtyard Marriott's extranet provides marketing reports and operating manuals to all its operators throughout the United States.
- An extranet setup by a pharmaceutical manufacturer allows hospitals, pharmacies, retail stores, and clinics access to an order processing and inventory tracking system, thus providing on-line access to its provisioning schedules.

The extranet network and security components are similar to the ones described in the intranet section. What does change is the context of their application. For example, the security policies and access control rules will have different context when applied to an extranet but the physical devices and other components do not change. We will further explore these components in Chapter 5.

1.6 Conclusion

In this chapter we introduced the concepts and definitions of the Internet, intranet, and extranet. We first distinguished between the terms *Internet* and *Internet technologies*, and then saw a glimpse of how the Internet technologies are applied to an intranet or an extranet setup. It is quite apparent that to set up an effective intranet or an extranet, you need to have a clear understanding of the network and security requirements. In the next three chapters we will explore the network and security components in more detail including the available technology choices, their selection criteria, and their pros and cons.

2

Wide Area Network Components

A typical intranet implementation will have the following WAN components:

- A physical connection to the Internet;
- A physical connection to a frame relay cloud;
- A WAN router and network addressing scheme;
- A configuration of frame-relay PVC on a WAN router;
- A configuration of Internet-based VPN on a WAN router or WAN firewall;
- A WAN network-routing protocol like border gateway protocol (BGP) for external connection, and an open shortest path first (OSPF) for an internal connection.

In this chapter we will look at these WAN components in detail. We will explore questions like:

- Which technology to use and why?
- Why certain routing protocols are preferred over others?
- What are the different WAN topologies and the criteria for selecting the best WAN topology, both from physical and routing perspective?
- How do you build a high-resiliency WAN infrastructure?

Figure 2.1 shows different technologies for connecting networks that are spread over wide geographical locations. The choices for WAN technology are:

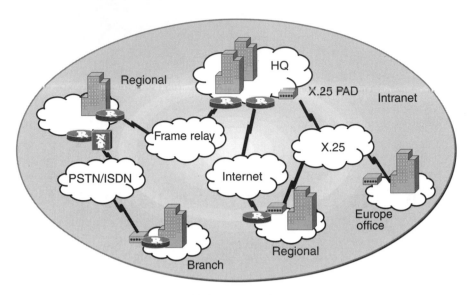

Figure 2.1 WAN technology options for intranets and extranets.

- Asynchronous dial-up connections on demand using PSTN;
- Dedicated digital point-to-point serial connections to an Internet service provider;
- Packet-switched technologies like X.25, frame relay, ATM, and switched multimegabit data service (SMDS);
- Circuit-switched technologies like ISDN.

Before we consider each one of these technologies in more detail, let's refresh our memories of the famous open systems interconnnection (OSI) model. This will allow us to map each one of these technologies on to the OSI model and understand the layer at which each one of these technologies operates. Figure 2.2 shows the OSI model (or the OSI stack).

In WAN technologies, the bottom three layers are of the most relevance. Figure 2.3 shows the various protocols used in these technologies at the physical, data link, and network layer of the OSI model. For example, X.25 operates at all three levels, while frame relay operates at layers 1 and 2 only. During the following discussion we will refer to Figure 2.3 and identify protocols used in each WAN technology implementation.

OSI Model

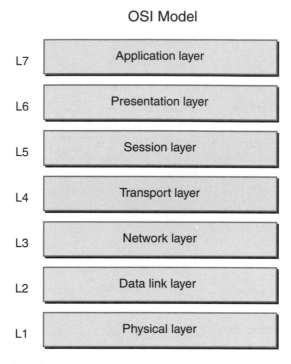

L7	Application layer
L6	Presentation layer
L5	Session layer
L4	Transport layer
L3	Network layer
L2	Data link layer
L1	Physical layer

Figure 2.2 OSI architecture.

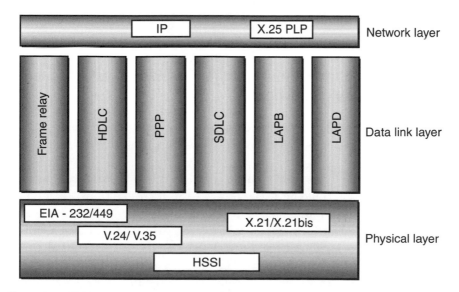

Figure 2.3 WAN technology mapping onto OSI.

2.1 Asynchronous Dial-Up Connections On-Demand Using the Public Switched Telephone Network

In this setup, the router is programmed to set up dial-up connections on demand to a remote location using the PSTN network. The LEC, like Bell Atlantic or a national telephone company, usually provides the PSTN network. The router at a branch location dials into an NAS at the regional office location, as shown in Figure 2.1. The two end points then enter into a point-to-point protocol negotiation phase with the router. Once the PPP negotiations are completed, the branch router can communicate with the regional-office router at the network layer. We will explore the PPP setup process in more detail later in this chapter. The router is usually programmed to disconnect the dial-up session after a certain period of inactivity.

Pros:

- This type of wide-area connectivity is the most cost-effective way of connecting low-traffic branch offices to the rest of the company.

- It is simple to set up.

Cons:

- The major limitation of such a setup is the low-bandwidth (up to 56 Kbps) connection. The bandwidth is suitable for low-traffic applications only and does not scale as demand for bandwidth grows.

2.2 Dedicated Digital Point-to-Point Serial Connection to an Internet Service Provider

In this setup, depending on the speed of the connection, various physical and data link layer technologies come into play. Please refer to Figure 2.4 for the rest of this discussion.

Figure 2.4 shows a typical setup for a point-to-point serial connection. The serial connection could be a fractional T1 (56 K to 512 Kbps), T1 (1.544 Mbps), or a DS3 (45 Mbps) connection. When a customer orders an Internet connection from a local or national ISP, the ISP will work with an LEC or CLEC (competitive local exchange carrier) to bring the circuit physically to a customer location. This part of the circuit is shown as a local-loop connection in the figure. The ISP then connects its network to the LEC network at one of its POPs. The line (or circuit) charge to the customer normally includes:

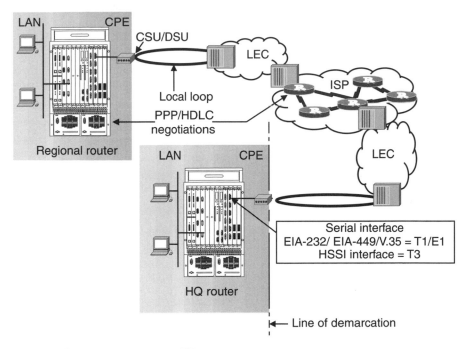

Figure 2.4 Serial connection to an ISP.

- The local-loop charges, which are distance sensitive. These charges are either bundled with the ISP charges or are billed separately by the LEC.
- Charges for the circuit between the LEC central office and the ISP POP. ISPs typically do not charge for this circuit.
- The Internet port charge, which is a monthly recurring charge.
- The installation charge.

The "line-of-demarcation" separates areas of responsibility between the customer and the LEC or ISP. This line is also referred to as the user-to-network interface (UNI). Customers own the responsibility for any equipment to the left of the line. The LECs will have certain requirements on the CPE like the channel service unit/data service unit (CSU/DSU), selection of line encoding and framing. In many cases, the CPE equipment, like the router or inverse MUX or CSU/DSU, may be owned and managed by the ISP.

If the customer wants to buy a fractional T1 circuit of 256 Kbps, he or she gets a T1 circuit that is rate limited by software configuration to 256 Kbps. If the customer desires bandwidth between a T1 and a T3 connection, then multiple T1 circuits are combined to provide bandwidth in increments of T1. This technique is known as I-MUXing or inverse multiplexing. An example is shown in Figure 2.5.

The amount of raw speed that you can get out of the physical circuit depends on the line encoding and framing scheme used. For example, North America and Japan use alternate mark inversion (AMI) and binary 8-bit zero suppression (B8ZS) while Europeans use a binary 3-bit zero suppression (B3ZS) scheme. With an AMI encoding scheme, a T1 (1.544 Mbps) circuit will allow you to get 1.344 Mbps speed, while B8ZS encoding will allow you 1.536 Mbps. In the case of framing, North America and Japan use super frame (SF) or extended super frame (ESF).

Once the physical connection has been established, the ISP router and the CPE router enter into a PPP negotiation. This allows the two routers to negotiate data link level parameters and optionally authenticate each other before they start exchanging network layer packets like IP datagrams. To understand the intricacies of the PPP negotiations, let's take a look at the workings of PPP protocol.

2.2.1   Point-to-Point Protocol

The PPP protocol provides a standard method for transporting multiprotocol datagrams over point-to-point links. The PPP protocol is documented in RFC 1661. The protocol operates at layer 2 of the OSI model. The PPP protocol setup process consists of three main phases:

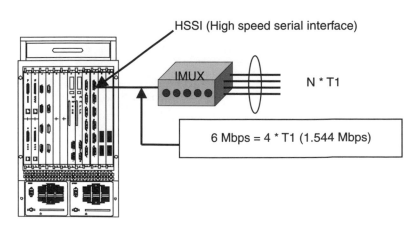

Figure 2.5 Inverse multiplexing for higher bandwidth.

- Link control phase (LCP);
- Authentication phase;
- Network control phase (NCP).

The PPP protocol is used for any point-to-point connection. This could be a dial-up or a dedicated connection. There is a slight difference in implementation and use of optional parameters while connecting using a dial-up connection, but otherwise the PPP setup process is the same. In the case of a dedicated connection the authentication phase is normally not implemented, since the arrangements between two parties are predefined.

2.2.2   PPP Operation

Link Control Phase

During the LCP phase, each end of the point-to-point link exchanges LCP configuration packets. These packets configure the link-level parameters such as:

- Maximum-receive unit (default *1,500 octets*) specifies the size of the packets that the end point can accept.
- Authentication protocol (default *none*) specifies the authentication protocol that will be used in the authentication phase before the NCP phase can begin. There are two types of authentication protocols, password authentication protocol (PAP) and challenge handshake authentication protocol (CHAP).
- Quality protocol (default *disabled*) allows peers to negotiate if the link should be monitored for quality. Only one value has been assigned to date for this option, "C025–link quality report."
- Magic number (default *zero*): this option allows the peers to detect loop-back links and other data link layer anomalies.
- Protocol field compression (PFC) (default *2*) allows peers to negotiate the compression of the PPP protocol field. This is useful on low-bandwidth links like a 14.4 Kbps dial-up link.
- Address and control field compression (ACFC) (default *both fields must be transmitted*) allows peers to negotiate the compression of the address and control field.

In most implementations the default values are used, thus simplifying the LCP negotiations phase.

Authentication Phase

By default, this phase is not mandatory. If a peer (in most cases an NAS or a router) requests for authentication, it must specify the authentication type, PAP or CHAP. We will take a detailed look at these authentication protocols in Chapter 5. The authentication phase is technically not a separate phase; it is part of the LCP phase.

Network Control Phase

The NCP phase begins only after the LCP phase has been completed. The NCP phase configures the parameters specific to the needs of the upper-layer network protocols that are being transported by PPP encapsulation. Some of the network layer protocol are IP, IPX, or AppleTalk. For example, in IP protocol, the configuration parameters include the setup of IP addresses of the end points.

This completes the data-link setup process. At this point we have the two end points communicating with each other at the physical and data-link level. Now the end points are ready to setup the network layer parameters.

The network layer parameters include the IP addresses of end points and IP header compression. The next step is to configure the routing protocols like RIP or BGP. If this is the only connection to the ISP, ISP might advise you to set up static routes advertising your networks to the ISP, which in turn advertises them over the Internet. If you have multiple connection to the same or different ISP, it is advisable to run a dynamic routing protocol like BGP to provide resiliency and load sharing. Figure 2.6 shows the configuration of a Cisco router connecting to an ISP with a single connection.

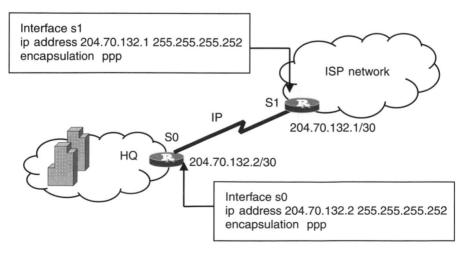

Figure 2.6 Configuration of a serial connection using Cisco routers.

Pros of point-to-point digital connection technologies:

- Simple to set up.
- High-bandwidth connections.
- Fewer line errors, hence better transmission throughput.

Cons of point-to-point digital connection technologies:

- Moderately expensive. Charges for a T1 circuit could range from $900 to $2,300 per month plus the local-loop charges. A T3, if available, could cost anywhere from $40,000 to $65,000 per month.
- Line activation can take a few weeks to a few months.

2.3 Packet-Switched Technologies Like X.25, Frame Relay, ATM, and SMDS

The packet-switched technologies deliver packets between two end points connected to a packet-switched network over a predefined path. The packets are referred to as "frames" in frame relay networks, or "cells" in ATM networks. The two end points are connected in a permanent (predefined) or a switched virtual point-to-point connection. For example, frame relay, X.25, and ATM have a concept of PVCs and SVCs. Figure 2.7 identifies the two different technology paths under packet-switching technology. The X.25 is the oldest of the protocols and is identified under the low-speed category, while frame relay, ATM, and SMDS are grouped under the new "fast packet-switching" category.

2.3.1 X.25

X.25 defines an interface specification for accessing an X.25 network. X.25 was one of the first packet-switching protocols and was introduced in the 1980s, when the transmission lines were nondigital and hence, error prone. To ensure reliable data transmission, X.25 incorporates network address, error detection, error correction, and flow-control mechanisms. Thus, when compared against the OSI model, X.25 is the only protocol specification that spans from layer 1 to layer 3. As shown in Figure 2.3, at the physical level the protocol specifies X.21/X.21bis standards, while at the data-link layer it specifies the LAP-B (link access procedure balanced) standard. The layer 3 standard is called the X.25 PLP (packet layer protocol). The packet layer defines the permanent or switched virtual path setup. Packets from various asynchronous datastreams are statistically

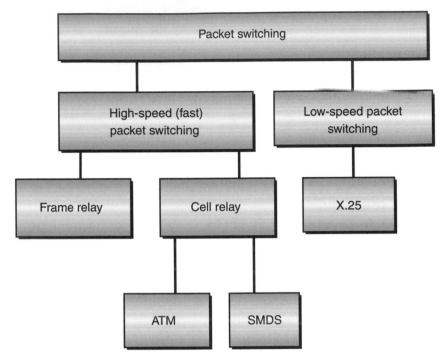

Figure 2.7 Packet-switching WAN technologies.

multiplexed on a physical circuit and sent synchronously across a packet-switched network. Packets travel across the X.25 network from switch to switch using the virtual path setup between the two end nodes. If a certain segment of the network is congested, the packets are queued on the switch (up to the maximum buffer value), until the congestion has subsided. If the number of packets in the queue exceed the buffer size, the packets are dropped and the end nodes then initiate the recovery procedures. As a result of the queuing mechanism, the packets encounter variable delay during the data transmission, making the X.25 protocol unsuitable for transporting time-sensitive applications like voice and video. X.25 can be used to transport network layer protocols like IP or IPX, by encapsulating them within X.25 packets. Each encapsulated network layer protocol is then identified by a tag. These encapsulation methods are specified in RFC 1356. Though X.25 is one of the oldest protocols, it continues to be a dominant protocol is Europe and Asia. The current implementations of X.25 networks can support speeds from 56 Kbps to a T1/E1. For more details on X.25 protocol and its operation, please refer to the Bibliography at the end of this book.

One of the fundamental assumptions during the design of the X.25 protocol was that the underlying transmission medium was unreliable and error prone. Naturally, these assumptions are not true in today's almost error-free digital and fiber-optic transmission environment. Hence, as the quality of the underlying transmission medium improved, new sets of protocols started replacing X.25. For example, frame relay, ATM, and SMDS. These protocols made an opposite assumption that the underlying transmission medium was error free and reliable, thus imposing fewer or no restrictions on error recovery and correction mechanisms within the protocol itself. This lead to the simplification of the packet-switching protocol and the associated packet-processing software on the switches, resulting in fast-packet switching and higher speed.

One of the next generation of packet-switching protocols is the frame relay protocol. The next section is devoted to the study of this protocol.

2.3.2 Frame Relay

Just like X.25, frame relay provides interface specifications for connecting a point-to-point circuit to a frame relay network. Frame relay maps onto layer 2 of the OSI model, as shown in Figure 2.3. Organizations could build their own frame relay networks or buy them from a service provider. Though you share the network with other users, it is the most cost-effective model, hence the most popular model as well. Since frame relay operates at layer 2, it can support transport of any network layer protocols like IP and IPX. RFC 1490 specifies the methodology of multiprotocol transport over frame relay.

Frame relay technology is well suited for bursty LAN applications running over a wide-area network. With the advent of public frame relay services, the cost of using frame relay as WAN connection technology has dropped dramatically. Frame relay technology also offers certain guarantees in throughput and reliability of service. In a recent survey by Frame Relay Forum for 1997, 46% of customers chose frame relay for WAN connection. Also the Frame Relay Forum estimates that 38,000 companies use half a million ports today.

Let's take a closer look at the incentives for implementing frame relay-based intranet.

Pros:

- It is cost effective.
- It is scalable up to certain number of sites (depending on the topology chosen).
- It is a stable technology that is available today.

Cons:

- Variable delays in transmission. This is an inhibiting factor for transport of time sensitive applications like voice and data.

- Traditionally, the frame relay circuits have been low-speed connections, from 56 Kbps to T1/E1 connections. Today, however, you can get high-speed circuits up to 45 Mbps. High-speed frame connections have been made possible by transporting frame relay over an ATM connection.

Design Considerations

Let's now look at the design considerations for selecting a frame relay circuit.

The key design parameters are number of PVCs, CIR for each PVC, and the resiliency parameters like a primary and a backup circuit. The backup circuit could also be a dial-back circuit.

The frame relay provides logical connectivity between any two locations by establishing PVCs between the two locations. Please refer to Figure 2.8.

The frame relay circuit from a service provider is terminated into a CSU/DSU at the customer premises. The CSU/DSU is then connected to a router serial port (e.g., CPE1 router at site A in Figure 2.8). Each serial port on a router can support multiple PVCs. Frame relay provides statistical multiplex-

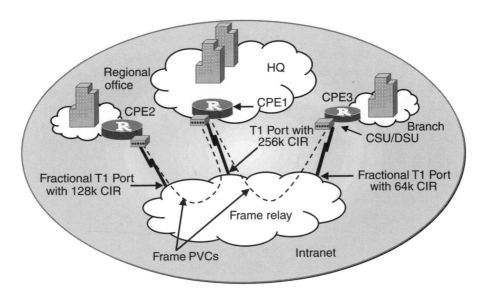

Figure 2.8 Frame relay network configuration parameters.

ing of multiple PVCs on a single access circuit. The PVC end points are configured as a subinterfaces on the router's serial port. We will take a detailed look at this configuration. In theory, each frame relay circuit and the associated terminating CPE router can support up to 1,024 PVCs, while in practice this number is limited to 100. Each frame relay PVC is configured with a CIR, which specifies the rate of guaranteed throughput for the PVC. The combined transmission speed of the multiple PVCs cannot exceed the frame relay port capacity. Thus, the design variables for a frame relay circuit can be specified as:

- Bandwidth capacity of the frame relay access port on the provider's frame relay network, specified in bits per second (bps);
- Number of PVCs using this frame relay access port;
- CIR parameters of each PVC;
- Specifications for the customer CPE device like a router or frame relay access device (FRAD).

For example, in Figure 2.8, the HQ has a T1 (1.544 Mbps) frame relay access port on the frame relay network. The CPE1 router terminates PVC1 (128K CIR) and PVC2 (64K CIR). The combined throughput requirement for the HQ site is 192 Kbps. As shown in the figure, the HQ site has been configured with a port capacity of T1 and a CIR value of 256K. This type of configuration can accommodate the current traffic requirements and also provides for a growth path for the future. The HQ site can technically burst up to full line rate of T1. Though the service provider provides traffic guarantees for the traffic meeting CIR value, any traffic above the CIR value is marked discard eligible and is transmitted on a best-effort basis.

Customers can implement various virtual network topologies like a hub-and-spoke configuration, partial mesh, or a fully meshed network configuration (See Figure 2.17 later in this chapter). We will consider some of these network topology options later in this chapter.

What Are the Criteria for Selecting a Frame Relay Service Provider?

On the surface, services that are offered by the frame relay service providers appear to be similar, although a careful inspection of the various service attributes reveals the true product differences. If you have a large number of sites (more than 15) you might consider issuing a request for proposal (RFP) to national or regional service providers.

What service level attributes are important while selecting a frame relay service provider? The following attribute list provides a good start:

1. *Port or connection speed:* Service providers offer port speeds in different increments (e.g., 8, 16, and 32 Kbps or 64, 128, and 256 Kbps). One must consider the base speed and the increment thereafter in case there is a future need for expansion in port capacity.

2. *CIR:* CIR values could range from 0 CIR to the maximum of port capacity. The CIR values can be assigned as simplex or duplex (i.e., if they are unidirectional or bidirectional). Check with the service provider before ordering.

3. *Usage charges:* Some service providers like MCI have usage-based pricing based on the number of bytes transferred. This is in addition to the monthly fixed cost.

4. *Availability:* This is the time between ordering of a circuit and its actual installation. This install time could vary from as low as 6 business days to 22 business days. If there is no port capacity available in a POP near your location, the service providers will pay the cost for a back-haul to a POP that has the desired port capacity.

5. *Modification time:* The time it takes to modify the port capacity or a CIR value based upon your request. Check if there are any SLAs that are offered by the provider in this area.

6. *SLA:* Many service providers offer SLAs in terms of maximum delay, number of packet discards, percentage availability of service, and so on. Due to intense competition in the frame relay market, service providers are offering attractive SLAs with certain money-back guarantees. All things being equal, your choice of provider will probably depend on this important attribute.

7. *CPE options:* The service providers sometimes bundle CPE devices like CSU/DSU, router or LAN switches in the pricing. The pricing could include the cost to buy or lease the devices.

8. *Management options:* The service could include optional performance reports and SNMP statistics in raw or report format. Most of the United States–based service providers are also working on on-line order entry and tracking systems.

9. *Enhanced services:* Confirm if the provider supports enhanced services like SNA over frame, X.25 over frame, IP over frame, or voice over frame. Depending on your requirements this could be one of the crucial attributes.

Factors in Selecting a Frame Router

Following is a list of design parameters that should be considered while specifying the frame router.

Router Processing Power in Packets-per-Second (PPS) and Router Memory

Requirements of higher processing power are driven by how much work a router has to perform to process an incoming or outgoing packet. If the router has to perform protocol conversion, process a routing table of 15,000 routes, and check on a lengthy access control list, it is going to need a lot of processing power. You can reduce the processing within a router by:

- Choosing static routes over dynamic routes. Static routes provide a simple and elegant configuration in most cases. But if you would like the router to react to change in the network environment, it is better to choose dynamic routing. For example, if there are multiple paths between two nodes and you would like the router to react immediately if the primary circuit gets disconnected, you should use dynamic routing.
- Reducing the access control list by a clever design of the filter list and by placing the most-hit filter rule on top.
- Reducing the number of routes in the routing table by creative combination of address blocks associated with various networks and subnets.

Available Number of LAN and WAN Interfaces and Expandability Options

At minimum, the router needs to have at least one serial interface for a WAN connection and a LAN interface for a local network. The type of WAN interface could be serial (up to a T1) or high-speed serial interface (HSSI) for speeds from 6 to 45 Mbps. Examples of various LAN interface types are ethernet (10 or 100 BaseT or gigabit ethernet), FDDI (100 Mbps), token ring (4 and 16 Mbps), or ATM (25 or 155 Mbps). The number of LAN and WAN ports should be expandable.

Protocols Supported by the Router Software

The router needs to support protocols like TCP/IP, Novell's IPX/SPX, DEC LAT, SNA, or Apple's AppleTalk over a LAN or a WAN connection. Routers also need to support various network routing protocols like X.25, SNA, RIP, OSPF, and BGP. In addition, if you are transporting one protocol type over the other, the router needs to support the ability to do protocol translation or encapsulation. For example, Motorola's MP6520 FRAD supports transport of X.25 or IP packets within a frame by directly encapsulating these packets within a frame. While in the case of Cisco routers, you will need to first encapsulate the X.25 packets within IP packets and then transport them over a frame relay network. This naturally adds the IP overhead and is not as efficient as the Motorola method of transport. If you are going to transport multiple protocols over your frame relay-based intranet, check support for RFC 1490 in the router software. As we mentioned before, RFC 1490 specifies multiprotocol transport over a frame relay network.

We will now take a look at two packet-switching technologies that are based on cell relay: ATM and SMDS.

2.3.3   Asynchronous Transfer Mode

Emerging interactive multimedia applications demand high-bandwidth, low-latency, reduced errors, and faster packet processing. Technologies like X.25 and frame relay are not geared towards supporting these highly bandwidth-demanding applications. A new technology is required that will accommodate the current and future needs of integrated audio, video, and data applications like telemedicine, multimedia conferencing, high-speed remote computing, high-speed LAN interconnections, and virtual reality. In addition, for interactive multimedia applications to be really useful, these applications need to be delivered directly to the user desktops. This necessitates a seamless transfer of the data from a user on one LAN to a user on a local or a remote LAN over a WAN. Today there is only one technology that spans from LAN to LAN and meets the demanding needs of the new breed of applications—ATM. ATM technology has matured in recent years and has found use in both LAN and WAN environments. With falling prices and increased market acceptance, ATM is making inroads into WAN and LAN networks. For example, you can buy a PCI interface card for PCs running Windows NT for as low as $200, while the price per OC-3 (155 Mbps) interface on an ATM switch has fallen to less than $1,000 as compared to $11,000 for an OC-3 port on a Cisco router and $14,000 for a DS3 (45 Mbps) port. Before we describe the origin and technical details of this wonderful technology, let's look at some of the pros and cons.

Pros:

- ATM provides scalable bandwidth. The bandwidth can be easily scaled from OC-3 to OC-12 (655 Mbps) to OC-48 (2.4 Gbps) to OC-192 (9.6 Gbps).
- Cost per port is falling.
- Extensive support for QOS parameters.
- Single technology for LAN and WAN.
- Fixed-delay and high-bandwidth features ideal for integrated voice, video and data applications.
- Multiprotocol support.

Cons:

- High level of complexity and resulting support costs.

This is the biggest disadvantage of implementing and running an ATM network. The current set of off-the-shelf network management and trouble-shooting tools are still not mature enough to handle large and complex installations. Today you will need to hire an expensive network engineering team to run an ATM network. This reason has also been responsible for limiting the deployment of ATM primarily to WANs, where it is absolutely needed. In the case of LANs, enterprises have opted to delay the deployment of ATM in favor of cheaper and simpler technologies like switched ethernet (10 Mbps), switched fast ethernet (100 Mbps), or fiber distributed data interface (FDDI).

Evolving ATM Standards and Vendor Interoperability

Since the standards are still evolving, vendors have opted to go ahead with their proprietary implementations, leading to interoperability problems. Fortunately for the industry, ATM forum has taken the lead in specifying standards like UNI 3.1, UNI 4.0, P-NNI 1.0, LANE 1.0/2.0, and MPOA, which should ease some of these issues.

Next, we will take a quick look at the ATM technology and then review some of the options available for WAN implementations.

When you compare an ATM stack to the OSI stack, you will observe that ATM works at the bottom-two layers, but the mapping is not very clean. Some of the functionalities of layer 3 are found in the layer 2 of the ATM stack. The ATM-to-OSI mapping is shown in Figure 2.9.

ATM Adaption Layer

The ATM adaption layer (AAL) is responsible for providing an interface between higher level services and underlying ATM layers. The AAL itself is divided into the convergence sublayer (CS) and segmentation and reassembly (SAR) sublayer. The CS provides an adaption functionality that depends on the type of higher layer application, audio, video, or data. The SAR layer at the transmitting end fragments the variable length frames or packets from higher layers into fixed size (48 byte) cells, while the SAR layer at the receiving end reassembles these cells, before passing them on to the higher layers.

The ATM Forum has defined various ATM adaption layers for adapting various types of services to the ATM transport. The ATM adapation layers are:

- *AAL1:* Specifies how TDM type constant bit rate (CBR) connection-oriented services are supported over an ATM network. This type of adaption layer is suitable for voice services over an ATM network.

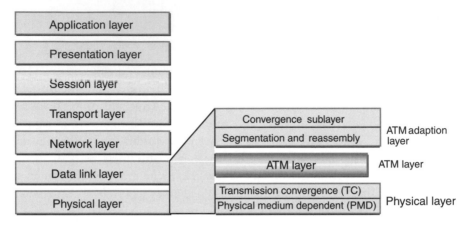

Figure 2.9 ATM layers.

- *AAL2:* Specifies how non-CBR type connection-oriented services are supported over an ATM network. This type of adaption layer is suitable for video services over an ATM network.

- *AAL3/4:* Specifies how variable bit rate (VBR) type connection-oriented or connectionless services are supported over an ATM network. This type of adaption layer is suitable for WAN services connecting bursty LAN applications over ATM. This layer is closely associated with SMDS over an ATM network.

- *AAL5:* Specifies how VBR-type connection-oriented or connectionless services are supported over an ATM network. This type of adaption layer is suitable for data services over an ATM network, especially IP services over ATM. This is a much-simplified version of the AAL3/4 layer and hence quite fast and efficient in processing higher layer protocol data units (PDUs).

ATM Layer

The ATM layer is responsible for establishing virtual connections and passing the ATM cells over these connections throughout the ATM network. The ATM layer is responsible for:

- Adding the 5-byte header to the 48-byte cell it has received from the ATM adaption layer, creating the 53-byte ATM cell.

- Managing the ATM virtual circuits (VCs) and virtual paths (VPs) and their remapping at each ATM switch. These virtual paths carry cells from various virtual circuits. A VCI and a VPI value uniquely identifies each cell in a virtual circuit.

- Additional functionalities like cell prioritization and flow control at the user-to-network-interface (UNI).

Physical Layer

The physical layer manages transmission of data bits over a physical medium. This layer is responsible for:

- Framing the data bits appropriate to the transmission medium. For example, the transmission medium could be synchronous optical network (SONET), synchronous digital hierarchy (SDH), DS3, E3, or FDDI.
- Providing some data link level functionality like checking for errors on cell headers and cell rate adaption, depending on the transmission system.

Due to the standardization of the data link functionality in the physical layer, ATM switches are now built with application-specific integrated circuits (ASICs) that provide this intelligence in the switch hardware, thus making ATM switching fast and reliable.

Now that we understand ATM layers and their functionality, let's see how ATM networks are built.

As an enterprise deploying this new technology, you are faced with the same old question, should you buy or build your ATM-based networks? The best answer is to use a combination strategy. Build your LANs and buy the WAN services from a service provider like MCI Worldcom or PSINet. Major ISPs in the United States have started to offer ATM connection services to their network at OC-3 (155 Mbps) or OC-12 (655 Mbps) speeds. The services are quite expensive and priced on a case-by-case basis. An ATM-based intranet would look like the intranet setup shown in Figure 2.10.

Figure 2.10 shows the LAN and WAN configuration with two major ATM interface specifications, UNI, and network-to-network interface (NNI). The UNI specification defines the connection protocol for connecting end-user devices to a private or public ATM switch, while the NNI specification defines the connection interface between any two ATM switches. Another interesting device is the ELAN module that allows connection of non-ATM–based LANs to an ATM network. We will cover the UNI specification, the ELAN module, and the associated protocols in the next chapter on LAN components of an intranet. In this chapter we will concentrate on the WAN components and the NNI specification.

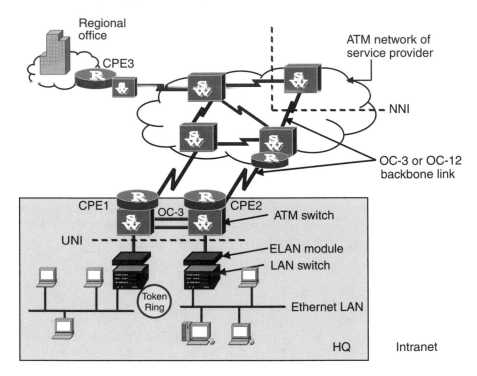

Figure 2.10 Intranet network architecture using ATM.

On the service provider side, most major ISPs like MCI Worldcom and Cable & Wireless use IP-over-ATM-over-SONET overlay model as shown in Figure 2.11.

In this model, the SONET ring provides the lowest level physical medium for the transport of ATM cells. The ATM switches are connected to each other using PVCs or SVCs at the ATM switching plane while a routed network running an IP routing protocol is overlaid on top of the ATM plane. That is, the router-based IP network uses the ATM network as a fast transport medium only and does not use any of the ATM routing protocols like private network-to-network interface (PNNI). Hence, the packets are routed at the IP level and not at the cell level. At this moment, none of the IP QOS parameters are mapped on to the ATM QOS parameters. This might change in the future, when routers use multiprotocol label switching (MPLS). ISPs use routing protocols like intermediate system-to-intermediate system (IS-IS) or OSPF over the backbone, while BGP is used for connecting customer networks to the ISP backbone. From a routing perspective, the ISP's backbone and the customer's network are considered to be separate routing domains or autonomous systems (AS).

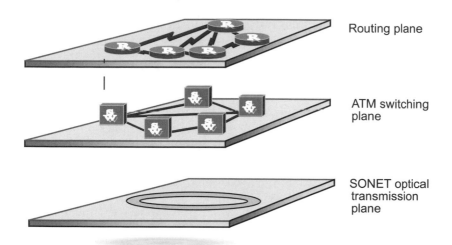

Routing plane

ATM switching
plane

SONET optical
transmission
plane

Figure 2.11 ATM in an ISP backbone.

When you get an ATM circuit from your network provider, it might be a

- Switch-to-switch connection;
- Switch-to-router connection;
- Router-to-router connection.

An ATM-based point-to-point connection from an ISP is most likely to be a router-to-router connection, instead of a switch-to-switch connection. This is because ISPs are most comfortable when they are offering a well-controlled access pipe to their network. The router-to-router connection allows ISPs to build protection mechanisms at the routing level between the two networks. The main objective is to provide a certain level of protection from misconfigurations on each other's network. An ATM switch-to-switch connectivity has the most promising setup for end-to-end QOS requirements; however, it is not preferred by ISPs since it is very difficult to manage and is prone to errors.

2.3.4 Switched Multimegabit Data Service

SMDS is another fast cell relay service, which is ideally suited for extending LAN connectivity at LAN speeds across a wide geographical area. It is a "fast packet relay" data service that is connectionless and uses cells of 53 bytes. SMDS was introduced as a stepping stone for ATM services. Though it was introduced around the same time as frame relay, it has quickly fallen behind in the United States as a technology of choice for service providers. Unlike in the

United States, SMDS has been very successful in Europe and its implementations far exceed the frame relay implementations. In the United States, SMDS service is currently offered by some of the regional Bell operating companies (RBOCs) and MCI. British Telecom and Deutsch Telecom are some of the SMDS service providers in Europe.

SMDS offers a connectionless service from speeds of DS1 to DS3. The SMDS cells can be transported over twisted pair, coax, or fiber-optic transmission medium. Its configuration is a star configuration with SMDS nodes attaching to a SMDS hub. These hubs are connected to each other to form an SMDS WAN implementation. One version of such implementation is called the metropolitan area network (MAN), shown in Figure 2.12.

SMDS implementation is based on the IEEE 802.6 DQDB (dual queue dual bus) standard. The interface between the CPE equipment (routers and SMDS DSU) and the SMDS switch is called service-to-network interface (SNI). The interface protocol is known as SMDS interface protocol (SIP). Data exchange interface, also known as DXI, was developed to ease the pain of transition to SMDS networks. With the DXI interface, the CPE router remains unchanged while the DSU performs the task of converting the incoming data units into 53-byte cells and passing them onto the SMDS switch.

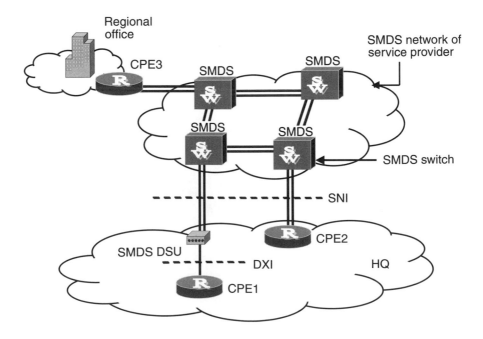

Figure 2.12 MAN using SMDS.

The bottom three layers of SMDS specifications map onto the first two layers of OSI stack. The three bottom layers are known as SIP1, SIP2, and SIP3. SMDS uses a 10-digit addressing scheme based on E.164 addressing format. The addresses resemble telephone numbers.

Like most LAN protocols, SMDS is a connectionless service. It also provides support for LAN multicast based on "group addressing." SMDS nodes registered with the group address receive a copy of the SMDS protocol data units. Implementation of such a multicast feature is quite complex in connection-oriented ATM networks.

Pros:

- High speed, LAN interconnect service for a MAN;
- Supports integrated voice, video, and data services;
- Provides a nice transition path to ATM;
- Supports security features like source address validation and address screening.

Cons:

- Not very popular with service providers in the United States, hence, lacks national appeal;
- More expensive than frame relay.

If this is a stepping stone to ATM, why not make a direct transition to ATM? With falling prices of ATM equipment, ATM services are poised to take off. This will avoid an extra transition step to ATM.

2.4 Integrated Services Digital Network

Unlike the packet-switched technology, ISDN is a circuit-switched technology. The signaling protocol in ISDN sets up a circuit-switched connection, just like in a telephone network. During the duration of the call, the resources of the circuit are dedicated to the connection. We explore the ISDN protocol and its implementation in this section (Figure 2.13).

ISDN provides a digital transport service over a PSTN network. Following are some of the pros and cons of ISDN service:

Pros:

- It provides an integrated transport for interactive multimedia applications using voice, video, and data.
- Uses the existing telephone infrastructure.
- The service protocols and interfaces have been standardized.
- Provides scalable bandwidth.
- It is universally available in Europe and is becoming widely available in the United States.
- It is ideal for a SOHO (small office/home office) or a branch office implementation.

Cons:

- ISDN is still expensive. Since ISDN is a circuit-switched technology, the tariff is based on the usage. If the usage is not properly monitored, the ISDN bill for a month could be as high as $3,000 to $5,000.
- Not all areas in United States have good error-free telephone transmission lines for implementing ISDN service. This results in long installation time for turning up an ISDN circuit.
- New technologies like ADSL, xDSL, or cable modem will give ISDN strong competition in terms of pricing and availability in the United States.

2.4.1   ISDN Physical Setup

The ISDN end-to-end setup consists of an ISDN switch at the service provider end and an ISDN end device (router, ISDN phone) with an ISDN terminal adapter (TA) at the user end. In the United States, the network terminating equipment "NT1" or "NT2" is based on the customer's premises, while in Europe, the NT1 is part of a service provider network. An example of an ISDN router is the Cisco 700/800 series router. Cisco 700 series ISDN routers incorporate the NT1 unit and associated U, S, and T interfaces within the router. Figure 2.14 shows the back panel interface of a Cisco 766 router. The router also incorporates two analog phone connections, shown as a "R" interface in Figure 2.13, and a 10BaseT ethernet interface for local LAN connection. The Cisco 766/801 type of device is ideal for a home office or a small branch office. The router looks like an oversized external modem based on today's standard size for an external modem.

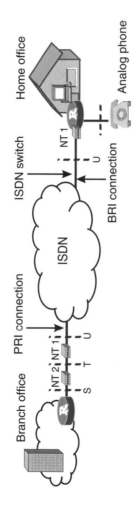

Figure 2.13 ISDN interface definitions.

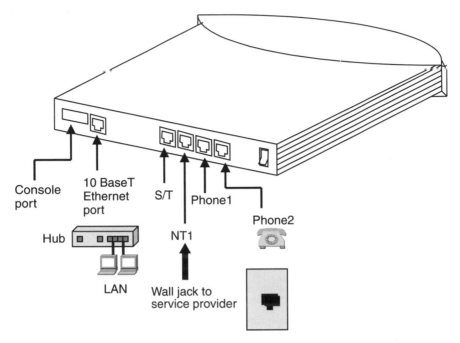

Figure 2.14 Cisco 766 ISDN router.

The ISDN interfaces shown in Figure 2.13 are defined below.

- *U:* The reference point between NT1 and the service provider switch;
- *T:* The reference point between NT1 and NT2;
- *S:* The reference point between an ISDN device and NT2;
- *R:* The reference point between the terminal adapter and non-ISDN equipment like an analog phone.

There are two types of ISDN services: basic rate interface (BRI) and primary rate interface (PRI).

2.4.2   Basic Rate Interface

BRI service consists of two bearer channels and one signaling channel, also known as 2B+D service. Each B channel provides 64-Kbps bandwidth and D channel provides 16-Kbps bandwidth. The total bandwidth provided by a BRI service is (2 × 64 + 16 + overhead = 192 Kbps). The two B channels can be used for data or voice traffic. The type of traffic is configured based on a service profile ID (SPID) assigned by the ISDN service provider. The D channel provides out-of-band signaling channel to setup an ISDN circuit-switched connection. The ISDN signaling protocol is called signaling system number 7 (SS7). We will take a look at SS7 later in this section.

2.4.3 Primary Rate Interface

PRI service consists of 23 bearer channels and 1 signaling channel, also known as 23B + D service. Each B channel provides 64-Kbps bandwidth and unlike BRI service, the D channel provides 64-Kbps bandwidth. The total bandwidth provided by a BRI service is (23 × 64 + 64 + overhead = 1.544 Mbps). The PRI service is offered over a standard T1 connection in the United States. In Europe, the PRI service consists of 30 B channels and one D channel. The total bandwidth is (30 × 64 + 64 + overhead = 2.048 Mbps). The PRI service is offered over an E1 connection. In both services, however, the D channel is primarily used as an out-of-band signaling channel; when it is not busy it can also be used as a data channel, providing an extra capacity.

2.4.4 Applications of ISDN

As shown in Figure 2.13, ISDN can be used to connect a branch or a home office through an ISDN network. ISDN is becoming a popular technology of choice for connecting telecommuting employees to the corporate LAN. Another

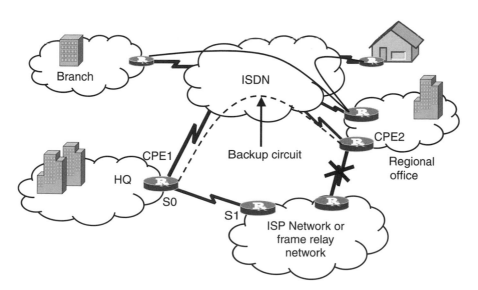

Figure 2.15 Use of ISDN in a dial-backup configuration.

growing use of ISDN is to act as a "dial-backup" circuit to a dedicated network connection. The dedicated network connection could be a frame relay connection or an Internet connection, as shown in Figure 2.15.

If the dedicated connection on a CPE2 router fails for some reason, the router would immediately bring-up the backup ISDN connection and reestab lish the connectivity to HQ. For this behavior, the ISDN connection and the dedicated connection must be on the same router. Another way to use the ISDN connection would be to use it as an overflow connection. If the dedicated circuit is experiencing congestion or is fully utilized, the CPE2 router can be configured to bring-up the ISDN connection for additional bandwidth.

An ISDN PRI connection can also be used as a channelized T1 circuit with 23 channels for PSTN call setup. Thus, if you are providing a remote access dial-in service to your mobile or telecommuting employees, you can buy a PRI line to your HQ and associate a single number for dial-in. All 23 channels of the PRI can be assigned to one hunt group and terminated on a network access server with digital modems. Examples of such network access servers are AS 5300 from Cisco and MAX TNT from Ascend. One such setup is shown in Figure 2.16.

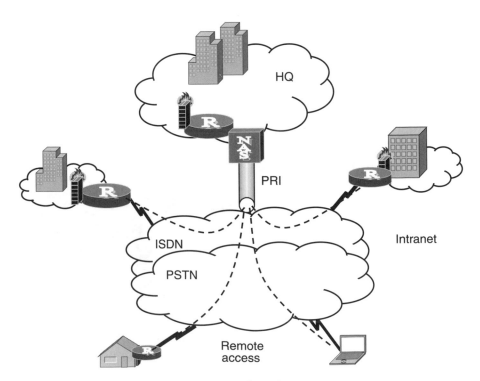

Figure 2.16 Use of ISDN in remote access configuration.

2.4.5 Security Features of ISDN

- Due to the digital nature of ISDN and extensive signaling setup, ISDN provides information on caller-ID and dialed number information (DNIS). Administrators at HQ can screen the incoming connection requests based on the caller ID and DNIS information.

- ISDN provides for "call-back." This is a great feature from a security, administrative, and billing aspect. It is a security feature because you know whom you are dialing to, while administratively it provides a central point of control. As for the call billing, since the HQ initiates the long duration calls, customers can get the biggest part of the bill from one location. This type of consolidation can be very handy while negotiating a price for the service from ISDN service provider.

- ISDN call setup can use many different higher layer encapsulation formats, like PPP, HDLC and X.25. The most common one is the PPP encapsulation. With the implementation of PPP, the NAS can request authentication via PAP or CHAP, adding additional level of security.

In our last topic on ISDN, we will briefly look at the signaling protocol for ISDN, SS7.

2.4.6 Signaling System Number 7

SS7 provides out-of-band signaling mechanism for telephone switches to set up an end-to-end circuit switched connection. The out-of-band signaling allows for full 64-Kbps bandwidth on the B channels. This setup is significantly different than other serial technologies where the signaling is in-band signaling, created by stealing bits from the data channel. For example, in a DS0 connection, this bit stealing reduces the effective data channel bandwidth to 56 Kbps instead of a full 64 Kbps.

SS7 is based on ITU specification Q.931.

2.5 WAN Topologies and Resiliency Considerations

In our discussions on various WAN technologies, we explored the inner workings of the technology. In order to complete our discussion, we need to consider the WAN deployment options regarding the network topology and the network resiliency. The cost of WAN circuits represents a significant portion of the networking budget. Any optimization in this department will have a direct impact on the company's bottom line. Another important aspect is the network resiliency. How resilient is your network architecture against failure of any one or multiple components? In today's networked world, uninterrupted connectivity is a basic need and is a fundamental mandate for a network engineering or management group. What is the best way to achieve network resiliency at reasonable cost? We will take a look at these issues in this section.

2.5.1   WAN Topologies

Irrespective of the WAN technology used, Internet, frame relay, X.25, or ATM, the WAN topology options can be summarized as follows:

- Full mesh;
- Partial mesh;
- Hub-and-spoke;
- Hierarchical in combination with above options.

Examples of these topology options are shown in Figures 2.17 and 2.18.

In the Internet-based network, the full mesh intranet or an extranet could be based on a full mesh of VPN tunnels. In the case of a packet-switched technology, it would mean a mesh implemented by permanent or switched virtual circuits. The full mesh design is by far the most resilient architecture, but it is also the most expensive and nonscalable architecture. The number of connections required to connect N number of nodes is a geometric progression represented by N (N-1)/2. For example, for a five-node network the number of connections would be 5 × (5-1)/2 = 10.

One way to reduce the cost and address scalability issue is to set up a partial mesh, with critical sites having a higher level of resiliency than the less critical ones. This is a classic trade-off between cost, resiliency, and scalability goals. Even further cost reduction can be expected with the availability of switched

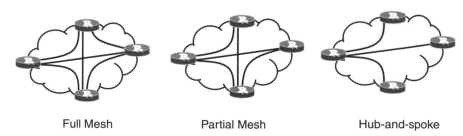

Full Mesh   Partial Mesh   Hub-and-spoke

Figure 2.17 WAN topologies for high resiliency.

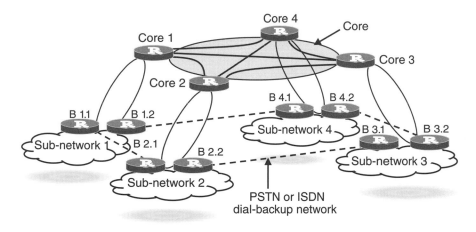

Figure 2.18 Hierarchical network configuration.

virtual circuits. The switched virtual circuits can be created on demand and provide a more scalable option.

As the number of nodes in a network grow, the best way to achieve scalability is to introduce a hierarchy in the network. The hierarchy could be logical or physical. The logical hierarchy is introduced by configuring routing domains that group together nodes that belong to a certain region, group, or city. Consider a hierarchical corporate network shown in Figure 2.18.

The four subnetworks could be representative of four regions, four cities, or four country operations. The subnetworks in turn communicate with their representative in the "core" network. For example, routers B2.1 and B2.2, interact with core 2 router. Routing protocols like OSPF or border gateway protocol version 4 (BGP4) support creation of routing domains like "areas" and an autonomous system to create routing hierarchies. The creation of such routing hierarchies enables us to:

- Create scalable networks. In this new network topology, the number of connections do not increase geometrically.

- Localize the network traffic and aggregate the routing updates. This leads to faster convergence of the network routes.

- Reduce the network impact area. If an upgrade is scheduled in subnetwork 2, the impact can be isolated to this subnetwork only.

- Improve fault isolation.

- Delegate network responsibility to local administrators.

- Increase security of the network, since any intrusion can be quickly isolated and restricted, before it spreads to other parts of the network.

Now that we have achieved scalability, let's take a look at the resiliency of the network. Please note that each core router in the network is a single point of failure. For example, if there is a failure in the core 2 router, it will isolate subnetwork 2 from the rest of the network. One simple solution to avoid this problem is to introduce a backup connection to another subnetwork using a dial-up connection. The dial-backup could be a 56-Kbps PSTN connection or an ISDN BRI circuit. Routers B2.1 or B2.2 can activate these backup circuits as soon as they detect a failure in a core 2 router.

Some other "low-tech" methods of achieving a level of resiliency would be to stock spares for the networking devices locally, including a spare router. Most networking vendors provide options for dual-power supplies or dual-route processing cards. These options can work in load sharing or redundant mode of operation.

In case you choose to order dual physical circuits for redundancy, make sure the local loops are divergent, thus protecting you from failure in any one of them. Similar considerations apply to power grids from the power company. If you can afford it, try to get separate power grids for your mission-critical data center.

2.6 Conclusion

In this chapter we have looked at the various WAN technologies to build your enterprise-wide intranet or extranet. We studied the inner workings of each technology, associated protocols, and their pros and cons. In the end, we covered various options for creating a highly scalable and resilient WAN infrastructure. In the next chapter we will focus on the LAN aspects of building an intranet or an extranet.

3

Local Area Network Components

A typical intranet or an extranet implementation will have various LANs connected by WAN links over a public or a private network. In previous chapters we explored the WAN network components, while in this chapter we will look at the LAN components in detail. We will explore questions like:

- Which LAN technology to use and why?
- What are the layer 2 and layer 3 switches? What is a VLAN?
- How to make use of ATM technology in an ethernet environment?
- Why certain routing protocols are preferred over others?
- What are the different LAN topologies and the criteria for selecting the best LAN topology?
- How do you build a high resiliency LAN infrastructure?

Over the last 20 years, LAN technology has evolved tremendously and now offers a variety of choices. The most commonly used technology options for implementing intranet or extranet LANs are:

- Ethernet;
- Fast ethernet (100BaseT);
- Gigabit ethernet;
- FDDI;
- Token ring;
- ATM.

These technologies can be broadly classified into three categories:

1. The contention-based ethernet technology;
2. The noncontention-based token passing technology (FDDI, token ring);
3. The connection-oriented ATM technology.

If we map the standards associated with each technology onto the OSI stack (see Figure 3.1), they specify the physical and the data link layer of the OSI model. For details on the ATM sublayer mappings please refer to Figure 2.9 in the previous chapter on WAN technologies for intranet and extranet.

In the next few sections, we will cover each of these LAN technologies, their pros and cons, and then finish with an example of typical LAN implementation using the technology under discussion.

3.1 Ethernet

Ethernet is the most popular LAN technology in use today. The term *ethernet* refers to any LAN technology that uses the carrier sense multiple access and collision detection (CSMA/CD) access protocol at the media access control (MAC) layer. There are different types of ethernet depending on their implementation speeds. For example, ethernet at 10 Mbps, fast ethernet at 100

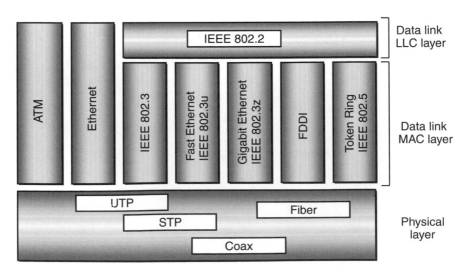

Figure 3.1 LAN technology options and OSI.

Mbps, or gigabit ethernet at 1,000 Mbps. In this section we will concentrate on the 10 Mbps ethernet implementation.

Ethernet technology was first invented at the Xerox Corporation in the early 1980s. The first implementation of the ethernet ran at 2.5 Mbps on a thick coax cable, which was later upgraded to 10 Mbps. The ethernet specification was then submitted to the Institute of Electrical and Electronics Engineers (IEEE) for standardization by an ethernet consortium consisting of Xerox, Digital, and Intel. The specification was standardized by IEEE with some modifications to the ethernet frame specification and was given the IEEE reference number 802.3.

The IEEE 802.3 specification specifies standards for the physical and data link layer. Based on different media types at the physical layer, the specification supports different transceiver types (see Table 3.1) like the unshielded twisted pair (UTP), shielded twisted pair (STP), coax, and fiber. The most popular media type today is the category 5 (cat5) UTP cable. The cat3 cables from the legacy environments are being replaced or augmented by cat5 cables. The great beauty of cat5 cable is that once installed, you can use the cable to run fast ethernet (100BaseT), FDDI (known as CDDI for copper cable-based specification), or ATM, thus offering a nice transition path without any rewiring of the offices or the network closets. For the rest of our discussion, when we refer to *ethernet* we are referring to the network using 10BaseT transceivers.

Table 3.1
Types of Ethernet Transceivers

Transceiver Type	Data Rate (Mbps)	Media	
		Segment Length	Type
10Base5	10	500	Thick coax
10Base2	10	185	Thin coax
10BaseT	10	100	Category 3/4 and 5 (UTP)
10BaseFL	10	2,000	Dual multimode fiber link
100Base-T4	100	500	Four-pair category 3/4 and 5 (UTP)
100Base-TX	100	500	Two-pair category 5 (UTP or STP)
100Base-FX	100	2,000	Dual multimode fiber link

Ethernet has found a wide popularity among network engineers and managers due to the simplicity of building ethernet LANs. A typical LAN implementation (see Figure 3.2) consists of an ethernet hub and a number of workstations connected by cat5 UTP cables in a "hub-and-spoke" or "star" arrangement. The logical arrangement looks like a bus architecture.

The cat5 cables, connecting each station to the hub, are "straight-through" cables, (i.e., the transmit and receive wires at each end of the cables are in the same pin position). Ethernet hub does the job of pin inversion. The transceivers on the workstation end are either built into the network interface card or are external (see Figure 3.3) to the workstation.

The simple network shown in Figure 3.2 can be expanded to accommodate growth in the network by adding additional hubs. The hubs can be stacked or connected in a cascade configuration to provide additional ethernet ports. Each hub or a collection of stacked hubs identifies a network broadcast domain, also called a network segment. As the number of workstations on an ethernet segment increases, the effective throughput of the 10BaseT LAN decreases. This decrease is due to the contentious nature of the CSMA/CD protocol. The CSMA/CD protocol works on the principle of "Send if the network is quiet; if you hear a collision of your transmitted packet, try sending it later after a random time interval, and keep trying until you succeed." If the network domain becomes too big or has a lot of workstations, the number of collisions on the network will increase, resulting in reduced throughput. The logical next step is

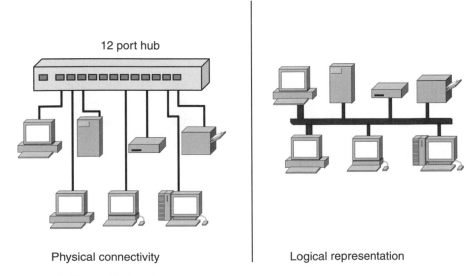

12 port hub

Physical connectivity   Logical representation

Figure 3.2 Ethernet hub-and-spoke setup.

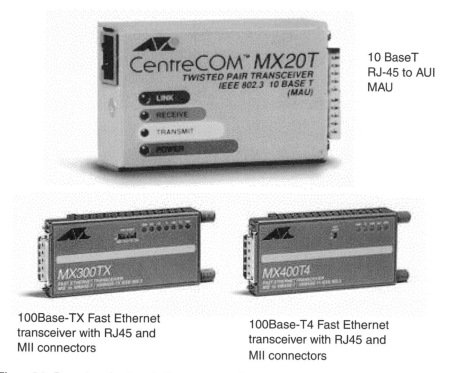

Figure 3.3 Examples of external ethernet transceivers.

to reduce the contention by creating smaller network segments, thus increasing the network throughput. You can create smaller network segments by using a bridge or a router (see Figure 3.4).

3.1.1 Bridges

Bridges serve a useful purpose for small LANs. If you want to extend a LAN or isolate the network traffic based on departmental grouping or a functional grouping, a bridge is a good choice. There are two major types of bridge technologies, *transparent bridging* and *source-route-bridging*. Transparent bridging is mainly used by ethernet networks, while the source-route-bridging is mainly used by the token ring–based networks. In transparent bridging, the bridge listens to the network traffic going from one segment to the other in a promiscuous mode and registers the source MAC address and the associated bridge port of the frame source in a bridge table (see Figure 3.4). If the bridge does not have the destination station's (STx.MAC, Port) entry in the bridge table, it broadcasts the frame on all bridge ports, except the originating port. The mapping table helps the bridge in passing the data layer frame to the right segment.

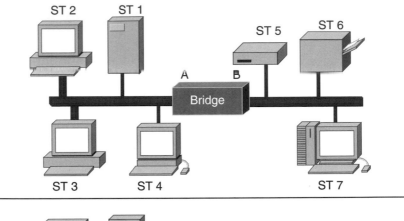

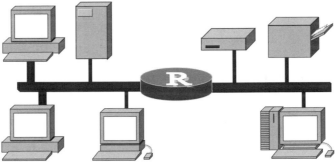

Figure 3.4 (a) Network segmentation using bridges and routers.

Transparent bridges isolate the intra-segment traffic and associated collision domain to the individual network segment, thus improving the network throughput.

As the number of segments and the number of nodes in a segment grow with complex network topologies, bridges do not scale well due to the possibility of network loops and unstable networks. The spanning tree algorithm was specially designed to prevent looping problems, but it still has scaling problems. What happens when you want to connect token ring–connected nodes with ethernet-connected nodes? How would you firewall traffic from one network segment to other network segment? The bridges were obviously not designed to these tasks, while some of these tasks are best accomplished at the network layer than the data link layer. The best device to accomplish these tasks is a router.

3.1.2 Routers

A router can address all the limitations of a bridge while adding a lot more functionality based on the network layer protocols. A router allows us to con-

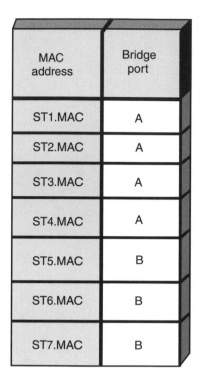

MAC address	Bridge port
ST1.MAC	A
ST2.MAC	A
ST3.MAC	A
ST4.MAC	A
ST5.MAC	B
ST6.MAC	B
ST7.MAC	B

Figure 3.4 (b) Bridge table.

nect two LANs spread over a floor, a building, a city, or a region using local or wide area links (see Figure 3.5). A router also provides services for protocol translation and encapsulation. A router can also act as an interconnection point between disparate LAN technologies. We have studied the WAN design and implementation techniques in the previous chapter. In the case of an intranet or an extranet implementation, a router can also provide firewall functionality.

In this section we have seen the workings of ethernet technology. We briefly looked at the role of bridges and routers in a LAN implementation. Next we will take a look at the fast ethernet technology.

3.2 Fast Ethernet (100BaseT)

With the advent of bandwidth-hungry client-server and multimedia applications, the effective bandwidth available on a 10BaseT ethernet network is not sufficient. Most of the ethernet segments will start to see a significant drop in effective throughput at 40% network utilization. This effective throughput is further reduced as the number of clients or servers on a segment goes up.

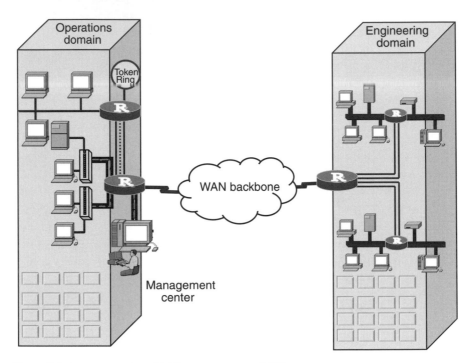

Figure 3.5 Interconnection of LAN segments over WAN.

The IEEE 802.3 Committee investigating the various approaches to increase the bandwidth of the standard 10BaseT ethernet implementation decided to increase the base frequency of operation from 10 to 100 MHz thus increasing the raw bandwidth to 100 Mbps. Apart from the speed improvement, what has really contributed to the overnight success of fast ethernet is that it is backward compatible with the ethernet (IEEE 802.3). The IEEE Committee, in its great moment of wisdom, decided to leave most of the features of ethernet intact, ensuring backward compatibility. Other competing standards, like 100VG-any LAN championed by Hewlett Packard, did not enjoy the same level of success, since the new standard was not backward compatible with IEEE 802.3. Though this standard has features that make it more desirable for multimedia applications, it does not use the same media-access protocol (CSMA/CD) like ethernet, thus making it incompatible with current implementation of ethernet.

Table 3.2 lists some of the key differences between ethernet and fast ethernet technologies.

Some of the significant differences, apart from the speed, are in the area of collision domain and flexibility of speed negotiation. Since fast ethernet still uses the CSMA/CD protocol, a reduced domain size allows the CSMA/CD

Table 3.2
Ethernet Versus Fast Ethernet

Features	Ethernet Type	
	10 BaseT	**100 BaseT**
Speed	10	100
Collision domain diameter (meters)	3,000	412
Autonegotiation of speed	Not supported	10 or 100 Mbps
Media-independent external interface	AUI	MII

protocol to work correctly on the same medium. The information about contention for a 64-byte packet at the farthest end of a domain must be able to travel back to the source end before the transmission of the next packet. With no change in the transmission medium, and since everything is happening 10 times faster, reducing the domain size was the only option for designers of fast ethernet. The fast ethernet also provides for a nice transition path for ethernet networks, by adding an autonegotiation phase to the new standard. The autonegotiation is implemented as a physical level signaling mechanism and allows the two end points, a station and a hub, to negotiate speed of transmission between 10 or 100 Mbps. The applications running on the station do not know the difference, but see a tremendous improvement in performance.

Fast ethernet uses different media types (see Table 3.1). The media types are classified in three categories: 100BaseTX, 100BaseT4, and 100BaseFX. The 100BaseTX uses the standard two-pair cat5 UTP or STP cable. If your network is already wired with this cable, your transition path is very simple since 100BaseT uses the same cable as 10BaseT. If you have legacy cat3/4 cables, then you will need four wire pairs to transmit the fast ethernet frames. The media independent interface (MII) adapter (see Figure 3.3) allows you to interface with any variety of the cabling infrastructure.

Fast ethernet can be introduced at a hub level as a high-speed replacement for ethernet hub or as an ethernet switch. What is an ethernet switch?

3.2.1 Ethernet Switch (Layer 2 Switch)

An ethernet switch is a layer 2 switch or a multiport bridge. An *ethernet switch* is a marketing term coined by the equipment vendors for revitalizing the age-old bridge technology. Switches allow us to introduce higher throughput for each end station by eliminating the contention and the resulting collisions in a traditional LAN backbone consisting of a hub. Figure 3.6 shows the differences

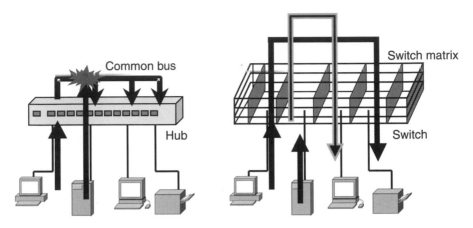

Figure 3.6 An ethernet switch versus a hub.

between operation of a traditional hub and a layer 2 switch. Switches add a tremendous amount of value by providing collision-free bandwidth to a user desktop or a server while still maintaining backward compatibility.

In the case of a hub, the frame signal travels on a common bus. The hub conditions and passes the incoming frame signal to all ports (network broadcast), except the incoming port. If another station simultaneously transmitted its frame onto the common bus, there is a signal collision. This results in all stations backing off for a random time interval before attempting to transmit again. As the number of users on the network grows, so do the chances for frame collisions, resulting in reduced network throughput. On the contrary, a switch does not have this collision problem since the frames are switched through a cross-bar, nonblocking switching matrix to the destination port. Unlike a hub operation, the packets are not copied to all the ports, thus avoiding any collision. The elimination of the network contention increases the network throughput. Each station connected to a switch can now transmit its packet without any fear of collision, resulting in full utilization of its 10 or 100 Mbps connectivity to the LAN backbone switch.

3.2.2 Switch Operation

Though we have described switches as multiport bridges, they are much more intelligent than bridges and come with quite a few management features. There are two types of switching technologies:

1. Store-and-forward switching;
2. Cut-through switching.

Store-and-Forward Switching

A store-and-forward switch processes the incoming datastream in a store-and-forward manner as described below:

- The incoming packets received by the switch are stored in the switch memory after the switch performs a cyclic redundancy check (CRC) on the frame to check for corruption. If the switch discovers a bad CRC, the frame is discarded.

- The switch then examines the source MAC address of the incoming frame. If the address does not already exist in the switch address table, the address is added to the table, just as in a transparent bridge.

- If the destination address is in the address table, the frame is then switched through the switch matrix to the destination port buffer. Various switch manufacturers design multiqueue output buffers to provide quality of service (QOS) features. Manufacturers prefer output buffers to input buffers to avoid head-of-line blocking. Head-of-line blocking is explained later in this section.

- If the destination MAC address is not in the switch table, it will "flood" (broadcast) the packet to all switch ports. (We will shortly look at various ways to limit this flooding in a switch.)

- When the destination device responds, the MAC address and the switch port information are added to the switch table and the frame is delivered.

Cut-Through Switching

In cut-through switching, the switch does not wait for the arrival of a whole frame. The switch makes the switching decision based on the pre-emble and the header information contained in the arriving frame. Since the switch does not wait for the arrival of the entire frame and it takes time to run the CRC check on the frame, it naturally speeds up the switching operation in the switch matrix, providing better throughput. On the flip side, if the frame has bad CRC, it is not discovered until the frame reaches the destination device, where is dropped, resulting in recovery by higher layer protocols. This results in retransmissions and reduced throughput.

The fast ethernet switches also provide support for full-duplex operation on a switch port, resulting in increased throughput. For example, with full-duplex operation, the effective throughput on a 100BaseT connection increases to 200 Mbps.

Switch Selection

When you want to select a layer 2 switch, the following parameters should be carefully considered before making the buying decision (see Table 3.3):

- Switch processing power;
- The architecture of the switch backplane, whether it has a single bus or a cross-bar switching matrix;
- The bandwidth of switching backplane or the system bus;
- Size of the main memory buffer;
- Whether the switch implements input queues or output queues;
- Whether the switch prevents head-of-line blocking.

Note   The head-of-line blocking is normally an issue with switches implementing input queue buffers, instead of output port buffers (see Figure 3.7). If a switch port has inputs from station A and B, and station A is generating a large amount of data, the frames from B are queued behind frames from A. The frames cannot reach the switch memory until all the A frames have been delivered to the switch memory. This phenomenon is called head-of-line blocking. This type of blocking is prevented by output buffers instead of input buffers. Since the switch has a nonblocking switching backplane, the frames are delivered to the output port without any switching delays.

- Does the switch provide support for any QOS mechanisms?
- Does the switch support full-duplex operation? Full-duplex operation doubles the effective throughput. This type of connection is very useful for switch uplinks or a high-volume server connection.
- How many MAC addresses are supported per switch port?
- Support for different LAN-media types. Modular switches can support multiple LAN media types as card slots in a chassis. For example, Cisco Catalyst 5x00 or 85xx series switches support 10BaseT, 100BaseT, mixed 10/100BaseT, gigabit ethernet, FDDI, ATM, LANE (LAN emulation), and route processor module.
- Another parameter to consider is the scalability of the switch chassis. How many ports of each type can it support? For example, Cisco 5500 switch can support a total of 528 ethernet ports, 264 10/100BaseT ethernet ports, or 32 ATM OC-3 ports.
- One of the resiliency considerations before deploying the ethernet switches in the field would be to add dual power supplies and any

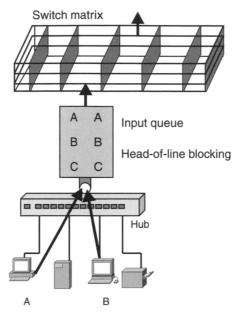

Figure 3.7 Head-of-line blocking in ethernet switches.

other resiliency features such as a dual CPU module. Check if the card slots are hot swappable. Once the switch goes in production, you do not want to bring all users down for a card maintenance.

- And of course, finally, the cost per port. Add yearly maintenance cost in the cost calculation.

A sample ready to use comparison chart for comparing various switch offerings is given in Table 3.3.

3.2.3 Virtual LANs

The case of a high-density switch "flooding" to all the ports is a waste of precious switch resources and does not contain the contention domain. One way to address this problem is to create virtual LANs, or VLANs. VLANs group the switch ports in a virtual LAN segment configuration. This restricts the segment broadcast to ports grouped under each VLAN configuration. Since the broadcast domain has been restricted to each VLAN, if you want to communicate between VLANs you have to go through a router. A router can then be configured with appropriate ACLs to either permit or deny the communication between two VLANs. This offers you an additional measure of security. LAN architecture with VLANs is shown in Figure 3.8. The two VLANs shown in the figure span multiple floor switches. This type of inter-switch VLAN is possible today

Table 3.3
RFI Questions for Ethernet Switches

Specifications	Manufacturers		
	X	Y	Z
Processing power			
Backplane architecture			
Backplane bus bandwidth			
Size of main memory			
Input and/or output queues?			
Head-of-line blocking?			
Any QOS mechanisms?			
Number of MAC addresses per port			
Media type supported Ethernet, 10/100BaseT, ATM, FDDI			
Scalability and port density			
Support for dual power supplies			
Total cost			

based on vendor proprietary solutions like Cisco's Inter Switch Link (ISL) protocol. The ISL protocol adds a VLAN specific ISL tag to every ethernet frame while communicating between two or more switches.

Figure 3.8 shows a high-resiliency intranet LAN architecture in a multistoried building. The two central switches, CWS1 and CWS2, are connected to each other by 100BaseT or a fiber connection. The central switches are connected to individual switches on each floor by a fiber or cat5 cable. These floor switches are then connected to a hub or a workgroup server with a 100 Mb connection. Individual clients are connected to the hub with a 10 or 100 Mb connection. Additional hubs and switches can be cascaded on each floor to give additional ports. The architecture implements a high-resiliency hot-standby routing protocol (HSRP) configuration. HSRP is a Cisco proprietary protocol.

3.2.4   Hot-Standby Routing Protocol

In an HSRP configuration, the two routers, R1 and R2, share a common MAC and an IP address (IP3). Each router is also configured with its own IP address, IP1 and IP2 (see Figure 3.8). One of the routers, let's say R1, acts as a primary router, while R2 acts as a hot-standby router. All the ethernet stations in the LAN point their default router address to the IP3 address. When both routers are up and running, R1 grabs the IP3 address and the associated MAC address

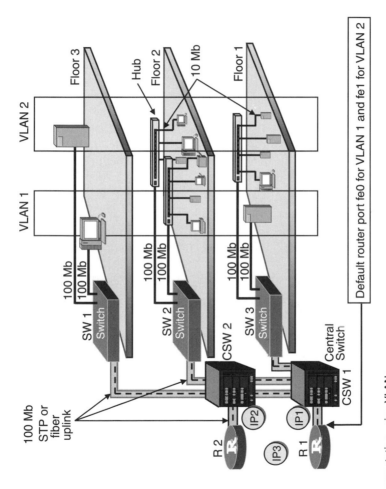

Figure 3.8 Network segmentation using VLANs.

and starts to route packets for all the LAN. Routers R1 and R2 exchange "hello" protocols to test if the other router is alive. If R2 detects a failure in R1, it immediately grabs the virtual IP address IP3 and the associated MAC address. The LAN stations do not know the switch has occurred, since their cached MAC address for IP3 has not changed. Once R1 comes back up, the switch over can be configured to happen automatically or manually.

With increasing demand for higher bandwidth, ethernet technology has evolved to support speeds in excess of 100 Mbps. The next evolution of ethernet technology is gigabit ethernet.

3.3   Gigabit Ethernet

The gigabit ethernet runs at 10 times the fast ethernet speed but uses the same CSMA/CD protocol. This makes it ideal for the next step in migration of fast-ethernet networks. The gigabit ethernet standard is defined in IEEE 802.3z specifications. The specification defines gigabit ethernet over fiber. The specifications for gigabit ethernet over copper have not yet been defined. The only major change from 100BaseT to 1000BaseT has been the increase in the minimum size of ethernet frame from 64 to 512 bytes. This increase has allowed gigabit ethernet to maintain the size of the collision domain at the same diameter as the fast ethernet.

The first ideal place for application of gigabit ethernet technology is in the LAN backbone. For example, in Figure 3.8 you can replace the 100 Mbps floor switch to central switch uplinks with gigabit ethernet uplinks. This higher level of aggregation means more stations on each floor with higher bandwidth. Currently, without gigabit ethernet, if you have to go above the 100 Mbps bandwidth limit you can set up full-duplex connections at 200 Mbps or turn to ATM at OC-3 (155 Mbps, full-duplex 310 Mbps) or OC-12 (655 Mbps, full-duplex 1.310 Gbps) speeds. But the introduction of ATM is not a simple process; the ATM to ethernet interaction needs a complex protocol such as LANE protocol and Multiprotocol over ATM (MPOA) to provide a routed ATM environment. This type of setup is complex and expensive to manage. Operation of gigabit ethernet today requires installation of fiber cables, however, it brings a tremendous amount of simplicity to the operation and maintenance of the network. If one has a current installation of FDDI network, the fiber installation can even be reused for gigabit ethernet.

Another ideal place for gigabit ethernet is the high-volume server connection to a switch. If your intranet or extranet server is conducting high-volume data exchange, like the exchange of CAD files, or ftp content, the bandwidth provided by gigabit ethernet would be ideal.

The biggest limitation to the full use of the gigabit ethernet bandwidth in today's LAN environment is the current PC bus architecture. The Pentium and Pentium Pro class machines cannot fully utilize the bandwidth provided by gigabit ethernet. New Pentium IIs with 100-MHz system bus and 64-bit PCI interface cards running at 65 MHz or higher are better suited to utilize the full power of gigabit ethernet.

3.4 Fiber Distributed Data Interface

In the days of 10 Mbps ethernet, 100 Mb FDDI has proven to be a high-bandwidth and high-resiliency solution of choice for a LAN-backbone technology. But now in the age of 100 Mb full-duplex ethernet running on copper and gigabit ethernet, this technology is quickly being relegated to a legacy standard. The high-resiliency features of FDDI, which heavily contributed to its success, are still unsurpassed in their simplicity and reliability by any other LAN technology. Since this technology is still being used in many enterprise networks, we will take a brief look at the operation of FDDI.

3.4.1 Infrastructure

FDDI is implemented as a dual-counter rotating-ring structure and uses a token-passing mechanism to pass data frames between any two stations on the ring (see Figure 3.9). The dual ring implementation has one primary and one secondary ring. The primary ring is always active, while the secondary ring is always in a standby mode. Each station on the ring is connected to both rings, or dual-connected. Each dual-connected FDDI interface has A and B ports. If there is a failure in any one of the stations or a ring segment, the ring "wraps" (see Figure 3.9), and the two counter-rotating rings connect to each other forming a single FDDI ring. This resiliency feature effectively isolates the problem area and keeps the rest of the ring functioning. For example, in the first scenario there is a failure in station III. The FDDI ring wraps on the FDDI stations II and IV, creating one ring and effectively isolating station III. In the second scenario, there is a fiber cut between stations II and III. The FDDI ring wraps on stations II and III, again effectively isolating the fiber cut and keeping the communication going in rest of the network. The FDDI specification can protect against one failure only. Multiple failures in the ring create multiple isolated ring segments (see Figure 3.9) as shown in the third example of multiple fiber cuts. In this scenario, after the ring wraps we are left with two isolated rings I–II and III–IV.

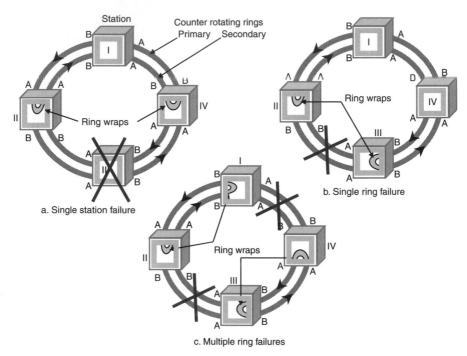

Figure 3.9 Failure scenarios in an FDDI ring architecture.

FDDI can be implemented on both fiber and copper cables. The copper specification is known as CDDI and supports 100 Mbps transport over twisted pair wiring (UTP and STP). The fiber installations are mostly multimode fibers and can support a ring spanning 62 miles in length. The maximum station count is 500 nodes, with each station separated by a maximum of 1.24 miles and connected by single-mode fiber installation.

FDDI maintains fairness among each station on the ring by implementing target token rotation timer (TTRT). At the time of ring initialization, each station agrees upon the token holding time (THT). THT is derived from the TTRT. Any station can hold the token for THT duration and pass the data frames into the ring. Once the station THT has expired, the station has to release the token, thus allowing the downstream station to capture the token for its own data transmission.

FDDI specification specifies three types of FDDI connected hosts:

- SAS (single attached station);
- DAC (dual attached concentrator);
- DAS (dual attached station).

An SAS device connects to the FDDI ring through a DAC (see Figure 3.10). The stations on floor 2 and 3 are connected by a SAS connection to the FDDI concentrator. The DAC itself is connected to the FDDI backbone using dual connections. The main benefit of the DAC connection is that any failure or intentional power down of the end station does not affect the operation of the primary FDDI ring.

DAC is connected to both the primary and secondary ring and provides SAS ports for connecting end stations to the main ring. DAC also allows us to setup "dual-homed" stations (see Figure 3.10, i.e., stations that are connected to two concentrators with separate connections). This provides high-resiliency architecture and is used mainly for mission-critical devices like routers and servers.

DAS is a dual-attached device. It is attached to both the primary and the secondary ring. The router and the FDDI concentrators in Figure 3.10 are examples of how DAS is connected to both rings.

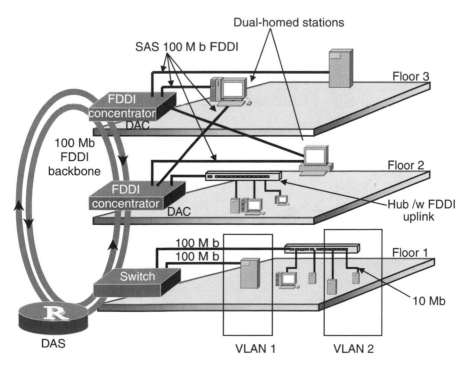

Figure 3.10 LAN backbone architecture using FDDI ring topology.

3.5   ATM in the LAN Environment

In the last chapter we have studied the use of ATM in the WAN environment. This section will focus on the use of ATM in a LAN environment. For a long time ATM did not find a widespread market acceptance due to its complexity, lack of need, and high cost. As the need started to be felt for high-speed technology and the prices for ATM equipment started to drop, ATM appear initially in the WAN and later in the LAN environment. Though the ball had started rolling for ATM implementations, there still remained a major stumbling block for widescale LAN deployment of technology. Since most of the LAN environments are populated by ethernet or token ring implementations, people needed a nice transition path that would take them from their current ethernet environment to the ATM environment. The ATM Forum (a consortium of ATM-related companies) came up with an interim solution that allowed the LAN traffic to be sent across an ATM backbone network. The solution was the ATM Forum specification for providing LAN emulation services over an ATM network. The specification is known as LANE. LANE allows emulation of a LAN-type (IEEE 802.3 or IEEE 802.5) environment over an ATM network. This task poses a great challenge since ATM is a connection-oriented technology while ethernet is a connectionless technology. How does LANE provide support for a connectionless technology like ethernet over connection-oriented ATM? We will answer this question later in this section.

Let's take a close look at the LANE architecture and its operation.

3.5.1   LANE Architecture and Operation

The LANE specification defines four main components (see Figure 3.11):

1. LES (LAN emulation server);
2. BUS (broadcast and unknown server);
3. LEC (LAN emulation client);
4. LECS (LAN emulation configuration server).

LAN Emulation Server (LES)

The LES implements the control-coordination function for an emulated LAN. The LESs provide a facility for registering and resolving unicast and multicast MAC addresses and/or route descriptors (token ring LANs) to ATM addresses, also known as the network service access points (NSAP) addresses. There is only one LES per ELAN. An LEC may register LAN destinations it represents and/or multicast MAC addresses it wishes to receive with its LES. LES provides

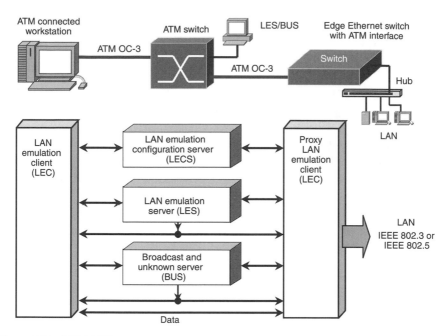

Figure 3.11 ATM LANE components.

answers to LEC queries for address resolution while resolving a MAC address and/or route descriptor to an ATM address. The LES may either respond directly to the LEC or forward the query to other clients for response.

Broadcast and Unknown Server (BUS)

ATM-connected end stations communicate with each other using dynamic virtual channel circuits (VCCs). To simulate a broadcast address LANE runs a BUS with a well-known or preconfigured ATM broadcast address. The BUS handles data sent by LECs to the broadcast MAC address ("FFFFFFFFFFFF"), multicast data, and initial unicast data sent by a LEC before the destination ATM address has been resolved. Each LEC is registered with one BUS per emulated LAN (ELAN). The main tasks of the BUS is to distribute data with multicast MAC addresses (e.g., group, broadcast, and functional addresses), to deliver initial unicast data, and to distribute data with explorer source routing information. The multicast packets are only sent to those LECs who have preregistered with their BUS at the time of initialization, indicating their interest in receiving multicast packets.

All ATM-connected devices set up a VCC to the BUS at the time of transmitting broadcast packets and the BUS sets up a point-to-multipoint VCC with each ATM-connected device for sending them the updates. Since all devices are connected to the BUS server, they can receive the broadcast packet just as in ethernet.

LAN Emulation Client (LEC)

The LEC performs data forwarding and address resolution functions and provides a MAC-level emulated ethernet/IEEE 802.3 or IEEE 802.5 service interface to higher level software. LEC provides these functions by implementing the LEC user-to-network-interface (LUNI) in order to communicate with other components within a single ELAN. Each LEC registers with both the LES and BUS associated with the ELAN it plans to join. Each LEC can be part of multiple ELANs. Since the VCCs are set up dynamically through the ATM network, the LEC does not need to be physically wired to each ELAN. An application of this great feature would be a backup server that needs to back up servers in multiple LAN segments.

Proxy LAN Emulation Client

The proxy LEC performs a "proxy" service for all the non-ATM-connected LAN stations planning to communicate over the ATM network. An example of proxy LEC would be an ethernet switch with an ATM uplink (see Figure 3.11).

LAN Emulation Configuration Server (LECS)

The LECS provides the initial configuration data for an LEC at the time of its initialization. The LECS configuration file is normally kept on the ATM switch or the ATM-connected router. LECS provides information like the available ELANs and the associated LES/BUS per ELAN. LEC uses this information to join an ELAN and register with the appropriate LES and BUS servers. An example of a LECS configuration file is shown in Figure 3.12.

In the configuration file, the first line specifies six ELANs, default, engineering, marketing, sales, support, and operations. The second line specifies a 20-byte ATM NSAP address; NSAP address of the LES server for each ELAN is specified next.

Note: NSAP Address   What does this long stream of numbers mean? The 20-byte NSAP address can be broken into subcomponents (see Figure 3.13).

If you observe closely the "lecs.cfg" file in Figure 3.12, the only thing different in each LES address is the last byte, also known as the selector byte. The selector byte for the "default" ELANs LES server address is 30. The selector byte of subsequent ELANs goes up in increments of two to 40 (3a). The first 19 bytes are common to all the NSAP addresses in this example. The 19 bytes denote the ATM connection of the server running the LES/BUS server (see Figure 3.11). The LECS configuration file can be used to denote a primary and a standby LES/BUS NSAP address per ELAN for a high-resiliency configuration.

'lecs.cfg file for a FORE Systems ATM Switch

Match.Ordering: default, Engineering, Marketing, Sales, Support, Operations
defaultAddress: 47000580ffe1000000f21c27080020480670c4.30
defaultAccept: XXXXXXXXXXXXXXXXXXXXXXXXXXXXXXXXXXXXXXXXXX
Engineering.Address: 47000580ffe1000000f21c27080020480670c432
Engineering.Accept: XXXXXXXXXXXXXXXXXXXXXXXXXXXXXXXXXXXXXXXXXXXXX
Marketing.Address: 47000580ffe1000000f21c27080020480670c434
Marketing.Accept: XXXXXXXXXXXXXXXXXXXXXXXXXXXXXXXXXXXXXXXXXXXX
Sales.Address: 47000580ffe1000000f21c27080020480670c436
Sales.Accept: XXXXXXXXXXXXXXXXXXXXXXXXXXXXXXXXXXXXXXXXXXXX
Support.Address: 47000580ffe1000000f21c27080020480670c438
Support.Accept: XXXXXXXXXXXXXXXXXXXXXXXXXXXXXXXXXXXXXXXXXXXX
Operations.Address: 47000580ffe1000000f21c27080020480670c43a
Operations.Accept: XXXXXXXXXXXXXXXXXXXXXXXXXXXXXXXXXXXXXXXXXXXX

Figure 3.12 Configuration example for LANE.

The LANE standard specifies four steps for setting an ELAN-to-ELAN communications:

1. Initialization;
2. Registration;
3. Address resolution;
4. Data transfer.

Initialization

As soon as the LEC is powered up it obtains its own NSAP address based on interim local management interface (ILMI) address registration.

The LEC then looks for the LECS to find its ELAN options and the address of the LES server associated with the ELAN it plans to join. LEC can obtain the LECS ATM address based on ILMI, via the "well-known" LECS ATM address or via PVC (0, 17).

Once the LEC has found the options for ELANs, it selects the ELAN of choice and communicates with the LES server for that ELAN.

Figure 3.13 Understanding NSAP addresses.

The LEC then declares if it wishes to receive address resolution requests for all the frames with unregistered destinations, receive specific multicast MAC addresses, or receive token ring explorer frames.

Registration

Once the LEC has communicated with the LES server, it registers the following information with the LES:

1. The list of unicast MAC addresses that the LEC represents;
2. The list of source route descriptors (i.e., segment/bridge pairs) that the LEC represents for source route bridging in token ring networks;
3. The list of multicast MAC addresses that the LEC will be receiving.

Address Resolution

At this point the LEC has joined the ELAN. The LES then:

4. Establishes a "control-distribute connection" to the LEC. The LEC and LES-initiated VCCs are used to communicate information such as the ARP request/response.
5. The LEC then sends an ARP request to the LES to obtain the NSAP address of the BUS server. The ARP request corresponds to the broadcast MAC address of "FFFFFFFFFFFF."
6. The BUS then responds to the LEC request by setting a "multicast-forward connection" with the LEC.

The LEC is now ready to transfer data.

Data Transfer

7. When the LEC receives a packet from the application layer that needs to be sent to another LEC or a proxy LEC, it sends an ARP request to the LES to identify the ATM address corresponding to the MAC address of the destination LEC.
8. The LEC also sends a broadcast request to the ELAN's BUS server, which promptly broadcasts it to all the LECs on the ELAN.

These two steps are redundant but are carried out to provide for a primary and a backup method to get the packet to the destination without delay. If the destination LEC is on the same ELAN, it will quickly respond to the broadcast from the BUS. Simultaneously, if the destination LEC is registered with the

LES, then the LES will respond with the NSAP address of the destination LEC. The requesting LEC will select whichever response comes first.

If the LES response is received first, LEC sends out a "flush" packet to the destination LEC via the BUS.

Once the destination LEC has acknowledged the receipt of the flush packet, the requesting LEC will start communicating directly with the destination LEC by sending packets to the destination NSAP address.

Just like an ARP cache in ethernet, each LEC maintains its own table of MAC to NSAP address in the cache. The cache entries are "aged out" over time.

The LAN architecture with ATM backbone and ELAN implementation is shown in Figure 3.14.

The ATM LAN backbone consists of dual OC-12 (655 Mbps) connected ATM switches with two routers and two edge devices connected to the ATM network with OC-3 (155 Mbps) connection. Every ATM connection is a full-duplex connection. The fiber is a multimode fiber. The edge devices are switch chassis with configurable-card slots. The right switch contains a switched ethernet and a token-ring module, while the right switch consists of switched 100BaseT module. The ethernet modules on both switches are configured with VLANs as shown in the figure. The ELAN module in each switch provides a

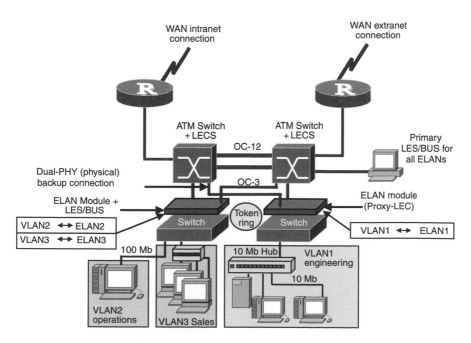

Figure 3.14 ATM LANE components.

proxy LEC functionality to connect the VLANs over the ATM network and is configured to map VLANs to ELANs. Both the ATM switches have been configured as LECS on the well-known ATM address. An external server has been configured as a primary LES/BUS server to offload the processing overhead to a dedicated machine. The ELAN module in the left switch acts as a backup LES/BUS server. Each router ATM interface acts as an LEC and participates in all three ELANs. The IP address on each router ELAN interface acts as a default route for communication between two VLANs. This completes our configuration of the network.

After the devices are turned on, the LECs go through the initialization, registration, address resolution, and data transfer process as described earlier. The dual-PHY connection is a standby physical connection that becomes operational if the ELAN module detects a failure in the primary uplink to the ATM switch. The routers can be configured in an HSRP configuration to provide added resiliency. An HSRP configuration protects against failure of any one of the routers, thus completing a high-resiliency intranet or an extranet LAN configuration based on ATM-backbone technology.

3.5.2   Multiprotocol Over ATM

One of the limitations of the previous design is the "single-arm" router connection to the ATM network. This single-arm connection splits the bandwidth available for router-to-ATM connection in half and could create a bottleneck for ELAN-to-ELAN (i.e., VLAN-to-VLAN) communication. One way to tackle this problem is to eliminate the router bottleneck by distributing the routing functionality into the ATM core and allowing creation of direct ELAN-to-ELAN communication, without going through the router. This is exactly what the MPOA standard accomplishes. The MPOA standard allows creation of a virtual router distributed across the network, using the edge devices for ports, ATM network for transport, and a route server for layer 3 routing functionality. The edge devices need to implement MPOA client software, while the route server needs to implement MPOA server software; the ATM switches themselves do not need anything specific to MPOA. Since MPOA runs over the LANE protocol, LANE implementation is a must.

The MPOA client provides a layer 2 and layer 3 forwarding functionality, while the MPOA server provides layer 3 address resolution (for example, IP to ATM address), layer 3 forwarding, and maintaining a layer 3 topology as well as reachability information with other route servers and routers.

Note: Tips for LAN Technology Migration to a Higher Bandwidth   How do you know when to upgrade? For a starter, user complaints on LAN performance is a good indication. But this is a reactive mode of operation; for your next promotion you need to be operating in a proactive mode. Here are some tips to excel in the proactive mode and stay one step ahead of the bandwidth-demand curve:

- The first step is to set up a baseline for the current performance characteristic of your network. If you plan to introduce new services, take time to capture the baseline. Without such a baseline characterization it will be difficult for you to quantify the performance impact of new services in the network.

- Analyze the current traffic patterns in the LAN environment. This will indicate which applications and which servers generate the most traffic on the network and at various times in a day or a week. It is also a good idea to maintain traffic analyzer probes in the network for daily or monthly reporting. The new probes from companies like Hewlett Packard and NetScout provide a secure Web interface and reporting functionality. We will explore these tools in more detail in the next chapter.

- The application and usage analysis will help you in creating departmental LAN segments to prevent traffic from a departmental application affecting nondepartmental users on the LAN. In switched environments, this will help you in creating VLANs in the ethernet environment and ELANs in the ATM LAN environment.

- It would be ideal to access some of the network simulation tools to understand the impact of new network services, but a general rule of thumb is close monitoring of the network can be equally effective in anticipating performance problems and taking appropriate corrective actions.

- If your current network backbone is a 10BaseT network and has 50% to 60% network utilization, an upgrade to a shared fast ethernet backbone would be in order.

- If you plan to introduce high-bandwidth services like interactive training in the intranet, with audio and video content, then based on your LAN utilization you might consider migrating the user population to either shared fast ethernet or switched 10BaseT environment. The latter option would be the most cost effective and least painful transition, since you do not have to upgrade the ethernet adapter cards in the user machines to fast ethernet. Of course, you have to plan the transition from hub to the switched environment at off-peak time.

- The next step would be to aggregate the switched 10BaseT connections on a 100BaseT uplink or a gigabit ethernet uplink.
- The bandwidth-hungry multimedia servers can be initially connected on a fast ethernet link to a shared 100BaseT backbone. Depending on usage statistics, this connection can be upgraded to a switched 100BaseT full-duplex connection.
- With falling prices of dual-speed NIC cards, any new procurement of LAN NIC cards should be of dual-speed 10/100BaseT type.

Having studied the fast ethernet and gigabit ethernet technology as well as ATM technology, a natural question to ask is, "Which technology is better suited for an intranet or extranet LAN backbone?" The answer is, of course, "It depends!" My recommendation would be that if your traffic type is mainly data and you want to upgrade the ethernet network without having to learn a new technology, fast ethernet and gigabit ethernet provide you with the best transition path.

ATM is the most suitable backbone technology if:

- Your traffic is a rich mix of audio, video, and data traffic.
- You need an incremental transition path for bandwidth with QOS features.
- You would like to make an investment into a single technology from LAN to WAN.

Presently, price per port for ATM is much lower than gigabit ethernet although you have to add the high cost of maintaining a qualified network engineering team well versed in the innerworkings of ATM technology. The cost could be worth it if you meet the criteria for ATM deployment.

Next we will look at the last of the LAN technologies that is likely to be used in today's intranet.

3.6   Token Ring

Token ring technology was developed by IBM as the main LAN technology for IBM LANs. It continues to play a major role in large corporations and government agencies where token ring is used to deliver mission-critical data in an IBM environment. There are approximately 20 million installations of token ring networks. The token ring standard is specified in the IEEE 802.5 specification and comes in two versions, a 4 Mbps version and a 16 Mbps version. An IBM token ring network is set up as a star network that logically represents a ring (see Figure 3.15). The token ring works on the principle of a token-passing network, just as in FDDI. If a station possesses a token it can transmit a data frame. Although both FDDI and token ring are examples of token-passing networks, their mode of operation is quite different.

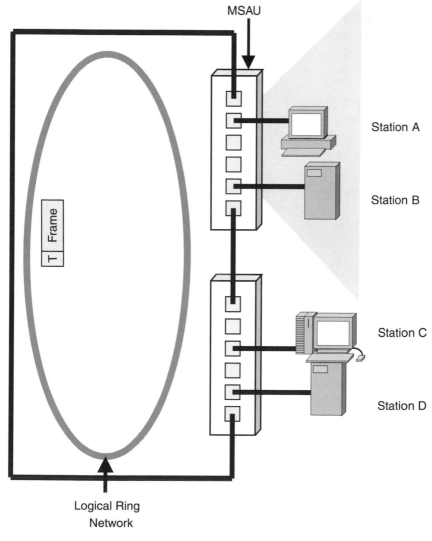

Figure 3.15 Token ring operation.

3.6.1   Token Ring Operation

Each station on a token ring network waits for a token to transmit its data frames. If station A gets the free token, it alters the bits in the token, appends its data frame to the token (now a start delimiter on the frame), and sends the whole sequence to the target station. The data frame is received and retransmitted by each intermediate station until the frame reaches the target station. The target station copies the frame, sets the copy bit in the frame header to 1, and retransmits the frame in the ring toward the originator. The originating station A grabs the returning frame, observes that the data was copied by the destination, and releases the token. The token is now available for other stations to transmit data. Since each ring has only one token, while the frame is traveling through the ring, there is no token available for other stations. Each station has to wait its turn until the token becomes available again. There is no contention between stations at any time. The token-passing network has the advantage of being deterministic. Since all parameters such as the number of stations and the ring size are known, both the network performance and the time taken to travel a ring are predictable. This feature is very important for time-sensitive applications like IBM's SNA network-based mainframe communication.

With the popularity of ethernet and its natural evolution to higher speeds, network managers have preferred ethernet over token ring. The war between these technologies has been won by ethernet. One of the main reasons for the decline of the token ring has been a lack of evolution strategy to higher speeds. Recently IBM has announced a formation of "high-speed token ring alliance" to address this issue in order to keep the installed base happy. The charter of this alliance is to first develop new specifications (in the IEEE 802.5 standards) for switched token ring running at 100 Mbps over copper and then develop the specification for fiber-based switched 100 Mbps token ring. Finally, a specification for a fiber-based token ring running at 1 Gbps will be developed. These specifications hope to stem the tide of migrating users from token ring to ethernet technology. At this point, this tide looks irreversible and it is expected that the share of token ring installations in the world will continue to shrink.

3.7   Layer 3 Switching

Until now, when we discussed switching, we always assumed it to be layer 2 switching. Now we will expand the term to include switching at layer 3, also known as layer 3 switching. Traditionally, switches perform the layer 2 switching and forwarding function, while the routers perform the network level (layer 3) routing function. Routers are general purpose computing devices with special routing software that performs the route selection and forwarding function.

Any software-based device will always have an inherent latency due to the software processing time and general-purpose nature of the computing device. This latency in processing packets at the layer 3 can be reduced dramatically if the routing software was embedded in a hardware module built using application specific integrated circuits (ASICs). The routing software instructions can now be executed at wire speed and in a distributed manner, increasing the performance of the whole layer 3 operation. This new methodology is usually referred to as layer 3 switching. (The marketing departments of the equipment vendors did such a fantastic job of hyping switching that the switching concept has now been elevated to the transport layer of the OSI model and is naturally called layer 4 switching.)

In addition to the performance aspects, layer 3 switching brings low-cost routing and advanced features like VLANs, RSVP, and MPLS to the edge devices. Since the decisions can now be made at the edges, the routing path is now optimized right at the ingress point of the network, resulting in improved performance.

The only limitation of layer 3 switching is that the current layer 3 switch implementations are limited to IP protocol only and that too with limited media options.

3.8 LAN Routing Protocols

The most popular LAN routing protocols are:

- Static;
- RIP (routing information protocol);
- IGRP (interior gateway routing protocol);
- EIGRP (enhanced interior gateway routing protocol);
- OSPF.

Other than the static routing, all other protocols can be broadly categorized as "distance vector routing protocol," "link state routing protocol," or a hybrid of the two categories (see Figure 3.16).

3.8.1 Static

Static routing is the simplest and most widely used routing technique in a LAN environment. Static routing is nothing but a table of routing destinations and the address of the next-hop router toward the destination. The table is configured

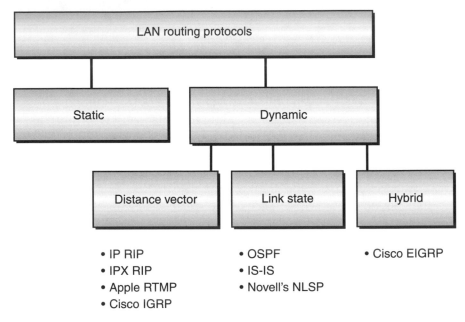

Figure 3.16 LAN routing protocols.

manually on all the LAN routers. Static routes also apply to all servers and clients on a LAN. These servers and clients are traditionally configured with a default gateway address. This default gateway address is nothing but a static route. For example, in Figure 3.17, each host VLAN2 and 3 is configured with a network address belonging to the respective class C address space and a default route pointing to the fast ethernet interface on router R3. The clients on the network can be configured manually or using a dynamic host configuration protocol (DHCP). We will study the DHCP protocol in the next chapter on service management.

Figure 3.17 shows an example of a typical intranet LAN environment.

The network has a dual FDDI backbone with four backbone routers. The routers R1 and R2 act as the border routers communicating with the rest of the intranet or an extranet router on customer/vendor/partner premises. The routers R3 and R4 act as the LAN firewall routers. The routers can implement router ACLs to filter incoming LAN traffic. The routers on the FDDI backbone could be configured to exchange routes based on RIP, IGRP, EIGRP, or OSPF protocols. Router R5 communicates with router R4 using EIGRP and is connected to a RIP network. The router R3 is connected to an ethernet switch. The VLAN segments configured on the ethernet switch extend to router R3 by inter switch link (ISL) protocol and virtual trunk protocol (VTP).

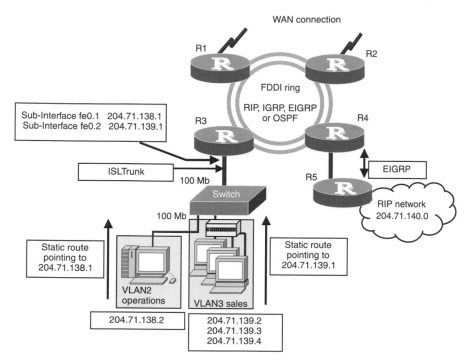

Figure 3.17 An example of intranet LAN architecture.

Note ISL and VTP are Cisco proprietary protocols. If you do not use Cisco devices, you can achieve the same results by an alternative setup shown in Figure 3.18.

While static routes are ideal for small networks, they do not scale well for large and complex networks. In the case of a complex network, if there is any change in the network topology (the change in the network topology could be due to a failure of one or multiple links, failure in one or more routers, or just plain upgrade of a network link), you would ideally like the routers to immediately communicate the change (route updates) and quickly settle on new routes for each source-destination network(s) pair, without any human intervention. This is exactly what the dynamic routing protocols do. Each dynamic routing protocol employs some routing algorithm that calculates the best path. Examples of such algorithms are the Bellman-Ford algorithm used by RIP and Dijkstra's Shortest-Path-First algorithm used by OSPF. Due to the dynamic nature of these protocols they scale very well for large and complex networks. The dynamic routing protocols can be categorized as:

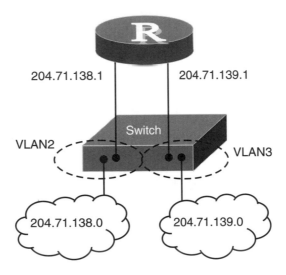

Figure 3.18 VLAN setup.

- Distance vector routing protocols;
- Link state routing protocols;
- A hybrid of these two categories.

3.8.2 Distance Vector Routing Protocol

The distance vector routing protocols calculate the routing distance (number of hops) and the best-route path (next hop) to each possible destination based on the information they receive from their neighboring routers. The distance vector protocols use a simple mechanism for providing routing information to their neighboring router: they provide their entire routing table. The routing information is supplied at periodic intervals or when triggered by events in the network. Since each router in the network needs to figure out nonlooping entries in its routing table before forwarding the table information, the convergence time (the time it takes to reach a consistent view of the network in all routers, after the exchange of routing information) is longer than the link state protocols. The slow convergence characteristics of distance vector protocols make it a second choice compared to the link state protocols which have a fast convergence time. We will study the link state protocol later in this chapter.

Examples of distance vector routing protocol are IP RIP, Novell's variation of RIP known as IPX RIP, Apple's own incarnation of RIP, RTMP (routing table maintenance protocol), and Cisco's IGRP. We will briefly look at each protocol and its pros and cons.

Routing Information Protocol

There are two versions of RIP protocol. The original RIP protocol is defined in RFC 1058 and the RIP version 2 is defined in RFC 1723. RIP is a distance-vector protocol and uses a simple metric (a metric is a variable that is used to distinguish the characteristics of various routing paths like distance and latency) of "hop count" to calculate the best route path from a source to a destination. The maximum hop count is 16, where the destination is declared unreachable. The routing updates are sent to neighbors every 30 seconds and when necessitated by network changes.

Pros:

- Simple to use.

- Universally implemented in routers and UNIX workstations. Though some of the implementations may be slightly different, they are interoperable.

- Well tested over the years.

Cons:

- The hop count of 16 was adequate when the networks were small. If your network has fewer than 16 hops, RIP is great, but if the network is growing very rapidly this could pose a serious limitation.

- The second major limitation of RIP is that it is a "classful protocol," (i.e., it understands only classful network boundaries like a class B or a class C network). In today's world of scarce IP addresses and use of variable length subnet masking (VLSM), this is a serious limitation. This limitation was addressed in the second version of the RIP protocol specified in RFC 1723. The RIPv2 supports VLSM.

- The third major limitation is the lack of variety in metric information. Since the only way for RIP to distinguish the path information is based on the hop count, it may result in suboptimal routing. For example, in the network shown in Figure 3.19, while communicating between R1 and R2 RIP chooses the suboptimal top path due to the lower hop count. Although the bottom path offers 10 times more bandwidth, RIP cannot make use of this information to select the bottom path.

- RIP also suffers from the fundamental limitations of distance vector protocols such as the slow convergence and periodic update of the entire routing table, which consumes the network bandwidth.

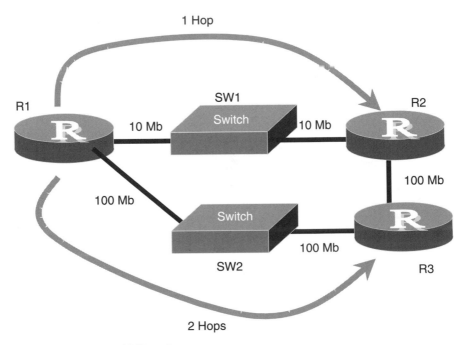

Figure 3.19 Suboptimal RIP routing.

Interior Gateway Routing Protocol

IGRP is Cisco's proprietary distance vector routing protocol. IGRP improved on most of the limitations of RIP. The biggest change was the removal of the hop count limitation. The IGRP metric uses a composite metric consisting of multiple links and network variables like bandwidth, delay, reliability, link loading, and maximum transmission unit (MTU) size. The new metric allows us to select the most optimal path based on the values of the metric parameters. Thus, in the previous example (see Figure 3.19), the router R1 would have chosen the 100 Mb path over the 10 Mb path. The improvements in Cisco's IGRP allow creation of highly scalable networks.

Pros:

- IGRP introduces concepts of "autonomous system" and "process domain," which allow creation of highly scalable hierarchical networks.
- Composite metric provides optimal routing.
- Flash update and hold-down timer mechanisms improve the convergence time.

- IGRP can maintain up to four paths to a destination, thus allowing route redundancy and traffic load-sharing capability.
- The periodic route update interval is 90 seconds resulting in a reduced number of updates in the network.

Cons:

- Like RIP, IGRP is a classful protocol and does not support VLSM. This limitation has been removed in the enhanced version of the IGRP protocol, EIGRP.
- IGRP is a Cisco proprietary protocol and is only supported on Cisco platforms.

Enhanced Interior Gateway Routing Protocol

Like its predecessor, EIGRP is a distance vector protocol, with distinct shades of a link state protocol. It is a Cisco proprietary protocol and uses same composite metrics as Cisco's IGRP protocol. Unlike IGRP, EIGRP uses diffusing update algorithm (DUAL) that works on the principle of diffusing the route information to a controlled set of routers in a controlled manner, resulting in fast convergence while preventing any occurrence of a loop.

Pros:

- Unlike typical distance vector protocols, the route updates are:
  - Not periodic, they are triggered by network changes only;
  - Partial, the routers participating in EIGRP only send changed routes not the entire routing table;
  - Bounded, the route updates are only sent to affected routers. All these features naturally conserve the network bandwidth and provide fast convergence.
- EIGRP supports VLSM and classless routing.
- EIGRP provides multiprotocol support and supports protocols like IPX and AppleTalk in addition to IP.
- EIGRP can be used to build scalable hierarchical network architecture.

Cons:

- EIGRP is more complex to troubleshoot than any of the distance vector protocols.

3.8.3   Link State Routing Protocol

Link state routing protocols are based on the concept of each node/router in the network having a complete map of the network. The route update messages, starting from an affected router, update the network information based on the change in the status of directly connected links. The receiving router copies this information and floods it to its neighboring routers. Each router then runs a "shortest path first" algorithm on the received information and calculates the shortest path to all other nodes in the network and creates a map of the entire network. The best routes to each destination are then inserted into the routing table. OSPF and Novell's NetWare Link-Services Protocol (NLSP) are examples of link state routing protocol.

Open Shortest Path First

OSPF has been developed by the Internet Engineering Task Force (IETF) and is specified in RFC 2328. The operational version of OSPF is version 2 and was first specified in RFC 1247. OSPF is a link state routing protocol and runs within a single autonomous system. Each OSPF router maintains an identical database describing the autonomous system's topology. From this database, a routing table is derived by constructing a shortest-path tree. In the event of any topological changes, such as a change in link status or failure in a router, OSPF quickly recalculates the routes by providing state updates (link state advertisements, or LSAs) to all neighboring routers, utilizing a minimum of routing protocol traffic. The neighboring router receives the update, copies it, and then floods the update to all its neighbors, thus propagating the information within a domain. OSPF provides support for equal-cost multipath routes thus allowing route redundancy and load-sharing capabilities. All OSPF routing protocol exchanges are authenticated, providing added level of security. The Internet Architecture Board (IAB) now recommends OSPF as the common IGP (interior gateway protocol) of choice (RFC 1371).

Pros:

- Since the link state updates are generated only if there is any change, the process is much more efficient than a distance vector protocol (this conserves the network bandwidth).
- Since each router floods the update information immediately upon copying, the neighboring routers do not have to wait long to receive the update, resulting in fast convergence. In the distance vector protocol, each router calculates a new route, updates the routing table, and then only sends the update to the neighbor. This causes delay and slow convergence.

- Each router has an accurate map of the network and can decide on the shortest path on its own (this results in optimal routing).
- Link state protocols support VLSM.
- Can build extremely scaleable hierarchical networks.
- OSPF supports complex composite metrics.
- Provides support for secure routing.
- It is easier to troubleshoot an OSPF network, since you can go to a single router and see which link in the network map is possibly broken.

Cons:

- Since each router needs to run SPF calculations every time it receives a link-state update, it creates a lot of CPU overhead.
- Depending on the size of the link state database, the router needs a lot of memory.
- Complex protocol to understand. (The RFC itself is more than 100 pages long! The book, entitled *OSPF: Anatomy of an Internet Routing Protocol*, by John T. Moy, author of the RFC, is the most authoritative book on OSPF and a must read if you are planing on deploying OSPF networks.)

The selection of the LAN-routing protocol is dependent on the scale and complexity of your network architecture. For example, if you have a four-node network, it would be an overkill to deploy OSPF in your network. Consider working with simple static routes, RIP or RIPv2, before jumping into OSPF. Other LAN-design goals like rapid convergence, high reliability, low overhead, traffic load, CPU utilization, ease-of-operation, and optimal routing will also dictate the selection of the routing protocol. A comparison chart of LAN routing protocols is given in Table 3.4.

3.8.4 LAN QOS

There are a number of ways to introduce differing qualities of service among the LAN traffic. Time-sensitive mission-critical applications like SNA applications or interactive voice and video over IP applications need special consideration while flowing through the LAN. If you have an ATM infrastructure, ATM QOS features implemented in ATM core and edge switches will be helpful in implementing traffic priorities. Even ethernet switches in the core or at the edges support multiple queuing mechanisms depending upon traffic types

Table 3.4
Comparison of Dynamic Routing Protocols

Protocol Features	Dynamic Routing Protocols				
	RIP	IGRP	EIGRP	OSPF	NLSP
Type	DV	DV	DV	LS	LS
Metric type	Hop count	Composite	Composite	Composite	Composite
Periodic update interval(s)	30	90	NA	NA	NA
Memory intensive?	No	No	No	Yes	Yes
CPU intensive?	No	No	No	Yes	Yes
Multiprotocol support?	No	No	Yes	No	No
Network architecture flat or hierarchical (H)?	Flat	H	H	H	H
Convergence speed?	Slow	Slow	Fast	Faster	Faster
Classless?	No	No	Yes	Yes	Yes
Proprietary?	No	Yes	Yes	No	Yes

and traffic flows. In today's environment, although various QOS mechanisms are gaining ground, the most common solution to a congestion or latency issue continues to be the brute force method of throwing bandwidth at the problem. In server farms, external traffic directors are used to provide load balancing and high availability of applications served by the server farm.

3.9 Conclusion

An intranet or an extranet LAN implementation has a wide range of choices for implementing scalable network architecture. The choice of the LAN technology and LAN-routing protocol will depend on the current technology implementations in your LAN, your area of expertise, the design goals for your next generation architecture, and of course your budget. In most scenarios a path of evolution rather than outright revolution is recommended.

4

Network and Service Management

Network and service management are two key components of intranet or ex-tranet deployment. These components provide the tools necessary to deliver and measure certain intra- or intercompany SLAs. These SLAs are measured on various QOS parameters such as availability, latency, and throughput. In this chapter we will take a look at the service parameters, available tools to capture the attributes of the service parameters, and the international standards these tools are based upon. We will cover the standards like simple network management protocol (SNMP), management information base (MIB), remote network monitoring management information base (RMON), desktop management force (DMTF), and common information model (CIM).

We will begin with the consideration of network management and then build the service management layer on top of the network management layer.

4.1 Network Management

Simply stated, network management provides configuration and monitoring of network devices like servers, routers, hubs, and applications running in the network. The best way to describe various components of network management is by describing the OSI FCAPS model.

4.1.1 OSI FCAPS Model

OSI was first to develop a framework for network management, known as the FCAPS model. The network management function was classified into five categories (see Table 4.1).

Table 4.1
FCAPS Model for Network Management

F	Fault management
C	Configuration management
A	Accounting
P	Performance management
S	Security management

- *(F) Fault management:* Provides detection, isolation, and correction of faults in the network operation.

- *(C) Configuration management:* Provides configuration changes like add, move, delete, and device-specific data collection including inventory management.

- *(A) Accounting management:* Provides collection of accounting data including accounting chargebacks based on the cost structure.

- *(P) Performance management:* Provides collection and analysis of the data traffic (e.g., link utilization).

- *(S) Security management:* Setup and reporting of various security mechanisms via thresholds and alarms.

SNMP was developed as an interim solution based on IP until a network management platform was developed using the OSI FCAPS architecture and the OSI transport protocol suite. Due to the slow development and deployment of OSI-based protocols, the success of the Internet, and the resulting deployment of IP, SNMP has found wide acceptance and is here to stay.

The first version of SNMP did not follow the OSI model and was limited to IP. Over the next two revisions, SNMP has edged towards the OSI model and is getting closer to the OSI-based network management protocol known as CMIP, or common management information protocol. SNMP will soon be in its third incarnation as SNMPv3. Table 4.2 shows the evolution of SNMP.

SNMPv3 has improved upon the previous versions, providing better performance and a solid security model. The various components of the SNMP architecture are specified by RFC 2271 through 2275 (see Table 4.3) and shown in Figure 4.1.

4.1.2 SNMP

An SNMP managed network consists of:

Table 4.2
Evolution of SNMP Standard

SNMP Version		RFC	Highlights
SNMPv1		1155, 1157, 1212	First SNMP implementation, used private and public community strings for security. Five operations: Get, Set, GetNext, Response, and Trap
SNMPv2	SNMPv2	1441–1452	Major improvements in performance with "GetBulk" and better security
	SNMPv2u	1909, 1910	Easier configuration
	SNMPv2*	IETF Draft	Remote configuration
	SNMPv2c	1901–1908	No remote configuration, uses community strings
SNMPv3		2271–2275	Takes most of the improvements from SNMPv2. Adds strong security and authentication model, support for remote agent configuration with SNMP, and unique ID for each SNMP engine

- A *management station,* also known as the *manager* (e.g., HP OpenView);
- *A managed device* with associated *agent* (e.g., a Cisco Router);
- The SNMP protocol for communication between the management station and the remote agent.

The SNMP model is a client-server model with the management station acting as a client querying the remote agent for information (see Figure 4.1).

The agent stores the management information regarding the remote device in a database known as the MIB. Each device manufacturer may provide a standard MIB and some MIB extension that is proprietary to the device. Just as one shoe size does not fit all, the standard MIB cannot provide device and manufacturer-specific information, hence the proprietary extensions. We will explore MIB groups in more detail later in the section. Each device manufacturer implements access routines as part of the agent software. The access routines gather the information from the managed device and then supply that information to the management station, upon request, via the MIB groups.

The management station consists of various applications and a graphical user interface (GUI) (see Figure 4.1). The GUI provides the most well-known representation of an SNMP system in everybody's mind. For most people, it is the network map or the flashing device symbol on the GUI that constitutes the

Table 4.3
SNMP Version 3 RFCs

SNMPv3	Highlights
RFC 2271	SNMPv3 architecture document
	• Defines a unique ID for each SNMP device, SNMPEngineID
	• Remote configuration of SNMP devices
RFC 2272	SNMPv3 message processing subsystem
RFC 2273	SNMPv3 applications subsystem
RFC 2274	SNMPv3 security subsystem includes the user-based security model (USM). USM provides:
	• User specific authentication for each SNMP packet
	• Message integrity with MD5 and protection against packet replay attack by way of time stamping each SNMP packet
	• Data privacy is provided by encrypting the packets using DES encryption.
RFC 2275	SNMPv3 access control subsystem describes the view-based access control model (VACM). VACM allows:
	• Each SNMP agent to enforce access policies
	• More granular control at the MIB object level

SNMP management system. But, as we are discovering, this observation is far from being true. The GUI component is a tiny portion of the whole SNMP management system. The other major applications resident in the management station are the event correlation engine and the reporting system. The event correlation engine correlates among the numerous alarms being received by the management station and tries to highlight the root cause. The correlation engine works on the logic programmed through the rule base. For example, if the management station starts to receive alarms from all the devices in the remote LAN, it is very likely that one of the aggregation devices, like a router or a hub, has failed, instead of all the devices failing at the same time. You can think of the rule base as a series of state engines for various scenarios. An example of a state engine is shown in Figure 4.2.

The reporting system consists of templates for various management reports such as the traffic statistics on a remote router or a graph of a trend based on the discrete sampling of a remote interface at five minute intervals. Enhanced reports are available based on the new extended MIBs known as the RMON MIBs. We will explore RMON later in this chapter.

Let's now take a detailed look at the concept of a device MIB.

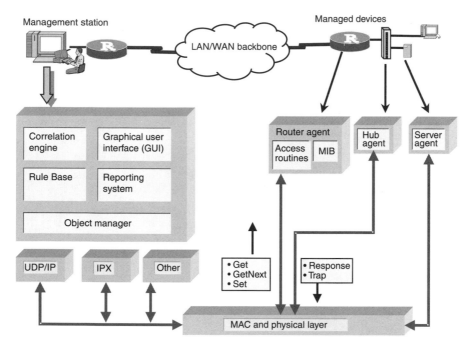

Figure 4.1 Typical SNMP-based network management architecture.

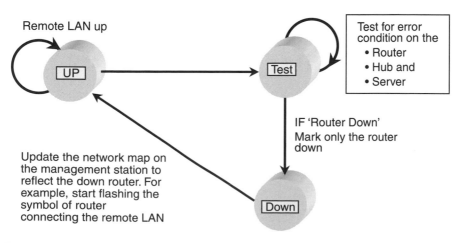

Figure 4.2 State engine for alarm correlation.

4.2 Management Information Base

An MIB is a device-specific database consisting of a number of management objects. Each management object represents a parameter of interest to the network manager, for example, traffic statistics on each device interface, or the number of active TCP connections on a device. Each management object is represented by an object ID (OID). Each object within an MIB is fully defined by a hierarchy of objects within an MIB tree (see Figure 4.3). For example, if an object in "standard MIB-2" is fully defined as ".1.3.6.1.2.1.2," a Cisco-specific MIB will have a prefix of ".1.3.6.1.4.1.9." The text equivalent of the Cisco OID would be, "iso.org.dod.internet.private.enterprise.cisco."

Each management object is part of a management information group. The first MIB standard, known as MIB-I (RFC 1156), consisted of eight information groups (see Table 4.4). The subsequent update to the MIB standard, MIB-II (RFC 1213), added three additional groups, shown in Table 4.4. Each vendor-specific MIB group is registered with an Internet standards body known as IANA (Internet assigned numbers authority), and is assigned a unique enterprise ID. Over 1,100 enterprise IDs have been assigned by IANA to date.

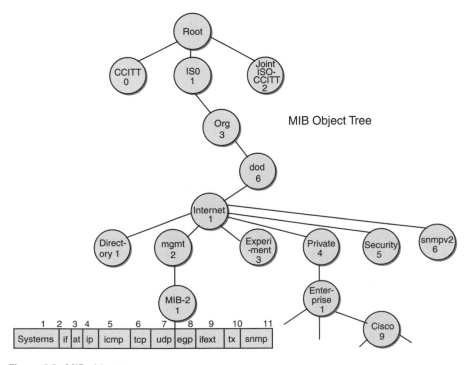

Figure 4.3 MIB object tree.

Table 4.4
Additional MIB Groups Defined by MIB-II Standards

MIB-I	MIB-II additions
System	Transmission
Interface	SNMP
Address translation	IF extensions
IP	
ICMP	
TCP	
UDP	
EGP	

Let's take a look at some of the MIB information groups.

- *Interface:* The interface group contains variables shown in Table 4.5.
- *Interface (IF) extension:* The interface extension group adds additional variables to the previous interface group. Additional variables include the following:
  - Keep a separate count of broadcast and multicast traffic;
  - Execute a test and then record the test results;
  - Keep a table of all physical addresses for which the interface will take traffic.
- *Address translation (AT):* The AT group keeps a complete map of all the next-hop IP addresses from the device and their associated MAC addresses. This group is now redundant due to the new variables added to the IP group that tracks the same information.

Table 4.5
Interface Group Attributes

Variable	Examples of Variable Value
Interface type and interface speed	Type = Ethernet type, speed = 10 Mbps
Interface status	Up or down
Incoming and outgoing interface traffic statistics	
Various error counts	

- *IP group:* The IP group contains variables such as:
  - Default TTL;
  - Reassembly time-out;
  - Traffic statistics of incoming and outgoing traffic from an interface;
  - Configuration table containing information on each interface, its IP address, subnet mask, and other IP parameters;
  - New next-hop IP address to the MAC address translation table;
  - IP routing table to get the next-hop routing information.
- *Transmission node:* This variable represents a node in the MIB definition and has several groups associated with different transmission technologies underneath. The examples of transmission technology groups are X.25, RS-232, FDDI, DS1, DS3.
- *SNMP group:* This group is a mandatory group for all MIBs and keeps track of information related to the operation of SNMP. For example, it keeps track of the number of SNMP commands issued to the device, the number of traps generated by the device, and the number and type of error encountered while trying to count MIB variables.

Note Once a vendor makes the MIBs available for its devices, the MIBs need to be compiled into the network management system (NMS). This is how the NMS knows about the management objects within device MIBs and can create queries specific to the MIB objects. If the vendor MIBs are written in conformance with the structured management information (SMI) specifications, any NMS compiler should be able to compile the MIBs. The first pass of the MIB compilation process converts the ASN.1 (abstract syntax notation number one) format to a standard programming language, usually C. A C language compiler on the NMS then compiles and links the MIB within the NMS.

4.2.1 Structured Management Information

SMI specifies how to use a subset of ASN.1 coding to define an information module. SMI specifications are used to create the standard MIBs. Vendors are also encouraged to create their device-specific MIBs using the SMI specifications. The MIBs conforming to the specifications are likely to be portable across multiple-management platforms.

4.3 SNMP Commands

The first version of SNMP introduced four SNMP commands (see Figure 4.1):

1. *Get:* Used by the NMS to retrieve information from the device MIB (e.g., get traffic statistics from an interface);
2. *Set:* Used by the NMS to set a parameter value in the MIB (e.g., set TTL value);
3. *GetNext:* Used by the NMS to get the next variable in the MIB table;
4. *Trap:* Used by the agent to inform the NMS of certain events (e.g., if the interfaces collision rate goes above the 2% threshold, then the router will generate a trap to the management station. How does the router know when to generate a trap? The manufacturer using agent-specific access routines programs this intelligence in the agent).

A "response" message is generated by the agent based on the "get" and "getnext" commands from the NMS. It is not an SNMP command.

SNMPv2 added the following commands:

- *GetBulk:* Used by NMS to invoke successive "getnext" commands to the MIB;
- *InformRequest:* Used for NMS-to-NMS communication. This command is valuable in setting distributed-management setup or providing NMS redundancy.

At this point we have looked at the various components of SNMP management architecture, including the management station, the remote agents, and the SNMP commands. In the IP domain, SNMP uses well-known UDP ports 161 and 162 for communication between management station and management agent. The network manager interacts with agents on port 161, while the agents send all the SNMP traps to the network manager on port 162.

Note If you have packet filters on routers or firewalls, make sure that the two UDP ports (161 and 162) are open for traffic.

If you plan to buy an NMS consider the features listed in Table 4.6 for a comparison of various NMS offerings from vendors.

4.3.1 SNMP Product Offerings

The network management marketplace has some undisputed leaders. The top four choices are discussed in the following sections.

Table 4.6
RFI/RFP Questions for Network Management System

NMS Features (SNMPv1, v2, or v3)	
Support for auto discover of the network	Check on a control knob to tune the depth of network discovery
Support for incorporating vendor MIBs and agents	This feature compliance varies to a large extent
Support for automapping of discovered nodes and the flexibility to rearrange the network and network links	Some NMS packages generate a cluttered map and make it very hard to customize the network view
Support for archiving of data	This archiving method could be simple flat files or an RDBMS system like Oracle
Runs on which operating system?	Sun Solaris, HP-UX, IBM AIX, or Microsoft NT
Is there a distributed management capability?	Usually support of a management hierarchy exists
Support for "autocorrelation"	This is very likely to be a third-party software or an expensive addition to NMS suite
Cost	This could range from $15,000 to $80,000

Hewlett Packard's HP OpenView Suite

HP OpenView Suite consists of many software packages including the SNMP management system known as the HP Network Node Manager (NNM) and HP IT/Operations. You can also combine the HP NetMetrix RMON management software with the HP NNM. The NNM 6.0 offers a Web-based SNMP management system. NNM also supports data archiving in a RDBMS system like Oracle. An autocorrelation engine is also available as an option within NNM. NNM also supports distributed network managers for fault tolerance and hierarchical management. The NNM is available for Sun Solaris, HP-UX, and Windows NT platform. For a Sun platform, it is recommended to have dual-processor UntraSparc servers with at least 128 MB RAM and at least a 4 GB hard drive. Similar specifications are recommended for an Intel-based system.

IBM Tivoli Group's NetView Network Management Suite

Tivoli's TME 10 NetView 5.x is available for Microsoft NT and UNIX platforms. Tivoli NetView management suite focuses on configuration, fault management, and performance management of the network infrastructure. The network management information is published using a Web interface. NetView also allows for grouping of network objects with common or similar characteristics. This grouping allows for easy change of views and policies on a group of network nodes. An event correlation engine allows you to filter traps and identify the root cause. The management data can be archived in variety of RDBMS platforms.

Sun's Solstice Domain Manager and Sun Enterprise Manager

Sun Solstice Domain manager includes the SunNet Manager, a network management system. Solstice Domain Manager in combination with Sun Site Manager provides a distributed management environment for a multisite operation. Both software packages are only available for the Sun Solaris OS running on a Sun Sparc/UltraSparc or X86/Pentium platform. While Sun Domain Manager is targeted for the enterprise operation, Sun Enterprise Manager is targeted for the telecommunications and service provider market. The Enterprise Manager uses Java technology.

Cabletron's Spectrum Enterprise Manager

The Spectrum Enterprise Manager emphasizes both the network and the service management aspects. The software runs on Microsoft NT and Sun Solaris platforms.

4.4 Remote Network Monitoring

The standard MIBs and the vendor MIBs play a significant role in the management of the network. The SNMP polling model of queries and responses works quite well in gathering a limited amount of data on a polling basis. But this model does not provide us the 24 × 7 monitoring of all the network segments. The SNMP model provides a snapshot of the network and does not provide a historical analysis of the network over a period of time. Wouldn't it be ideal if we could monitor the network remotely without overloading the network with the SNMP polls? How about gathering the data to analyze the network traffic at all seven layers of the OSI model? These were the thoughts behind the development of a new kind of MIB, known as the remote network monitoring management information base (RMON MIB). RMON MIBs provide information groups and MIB objects that can be used to gather information on all seven layers of the OSI model, while passively monitoring the network. (RMON allows information gathering on layer 1 and 2, while the RMON-II standards allow data gathering on layer 3 to 7.)

RMON MIBs are usually implemented in specialized network devices called network probes. Each network probe has its own CPU, memory, disk space, and out-of-band access or transmission port. Though stand-alone probes are quite popular, some device vendors have started implementing RMON MIBs directly in their network devices. For example, Cisco Catalyst 5xxx series implements a subset of the RMON MIB specification within the device itself. In most cases, the embedded RMON MIBs have less functionality than the dedicated hardware probes. The data gathered by the RMON MIB is invaluable in gathering network performance statistics and helping in problem diagnosis without intrusion in the network. We will take a look at various RMON-probe vendors and the RMON reporting software packages later in this section.

The first RMON specification was published in RFC 1271, which was later updated in RFC 1757. The RMON MIB provides MIB groups to capture the traffic statistics and network alarms. The specification deals with the traffic at the physical and the data link layer. The RMON-II specification adds additional information groups to capture information in the network layer and above.

The advantages of RMON probes implementing the RMON MIBs can be summarized as follows:

- RMON probes allow collection of network data and statistics, without involvement of the NMS. The probes can monitor various RMON variables for any exception condition, resulting in a "trap" from the probe to the NMS. This mode of operation replaces the network-intensive polling mode used in SNMP. This stand-alone operation is especially useful if the NMS system is isolated from the network segment by design or due to a network fault. The data gathered by the probe is then useful in conducting fault diagnostics.

- The probes can store data for historical analysis. This feature is extremely useful for identifying any network- or application-level trends. Fault diagnostics can be conducted based on the playback of the stored data.

- The probes can be configured to monitor various network parameters such as percentage of network collisions. If the percentage goes above a certain threshold (e.g., 2%), the probe is configured to generate a trap to the management station. This allows a proactive monitoring of the network with detail traffic statistics.

RMON probes also help gather additional information such as:

- Which protocol is the most used protocol on the network segment?
- What is the percentage distribution of each protocol?
- Which is the most used application?

- What is the percentage distribution of all applications?
- What pair of nodes is hogging the most bandwidth?

These statistics allow the organization to set up a baseline for the network and application services. The gathered data can also be used to conduct "what if" analysis, before allowing a new service on the network, (e.g., adding video-conferencing and intranet training (audio/video) service).

Let's now look at the MIB groups defined by RMON and RMON-II. Figure 4.4 shows the RMON MIB object tree. The OID of any RMON MIB group will have a prefix of "1.3.6.1.2.1.16."

The MIB groups defined by RMON MIB (RFC 1757) are:

- *Statistics:* Contains the statistics measured by the RMON probe for each of its monitored interface. This group as defined in the RFC is geared toward the ethernet network. New RFC 1513 is geared toward the token ring networks.

- *History control group:* Controls the periodic sampling of data from various types of networks. RFC 1757 also defines a new group called the "ether history group" that stores periodic samples of data from ethernet

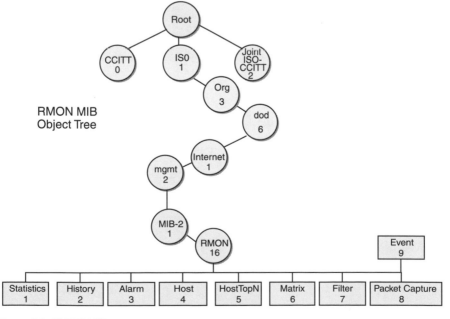

Figure 4.4 RMON MIB tree.

networks for historical analysis (e.g., the packet error count and the packet count).

- *Alarm:* Samples the network periodically and compares the samples against programmed thresholds. If the programmed thresholds are crossed (rising or falling), the RMON probe generates an event, which is processed as a trap to the management station.

- *Host:* The host group maintains statistics for each host discovered on the network. The group also maintains a table of MAC addresses associated with each host.

- *HostTopN:* The "hostTopN" group maintains a list of the top number of hosts based on a criterion programmed by the RMON management station. The traffic is sampled over a given time interval to determine the "hostTopN."

- *Matrix:* The matrix group reports on the number of conversations between a pair of addresses.

- *Filter:* The filter group provides filter for capturing selective data traffic.

- *Packet capture:* The packet capture group captures packets based on the filters provided by the filter group.

- *Event:* The event group controls the notification of events by SNMP traps.

A RMON device may or may not implement all the RMON groups. Implementation of all RMON groups is not mandatory, although if you implement any one of the RMON groups, you have to implement all the MIB objects under that group to claim compliance.

Note You have to be very careful in checking the claims of a manufacturer when they claim RMON compliance. Make sure that all the RMON groups of interest to you have been implemented. This could be a deciding factor when comparing vendors with equally good products. For example, the embedded RMON MIB in Cisco Catalyst 5xxx series implements only four MIB groups— the statistics group, the history group, the alarm group, and the event group.

4.4.1 RMON-II

The RMON-II standard (RFC 2021) provides new MIB groups for analysis of the network traffic from the network layer to the application layer. This addresses some of the limitations of RMON and provides valuable information on the network traffic. The new RMON-II MIB groups are:

- *Protocol directory:* Each RMON-II MIB group captures protocol types at multiple levels of the OSI hierarchy. The protocol directory keeps an inventory of all the protocol types it encounters and a description of the protocol types recorded. The probe can keep track of all the protocol types, although the recording could be limited to the protocol types it has been programmed to decode.

- *Protocol distribution:* The protocol distribution group keeps count of all the protocol types detected on the network.

- *Address mapping:* The address mapping group records all the MAC address-to-network address mappings and the interface on which they were last observed.

- *Network layer hosts:* The network layer hosts group captures and records the traffic originating from and terminating to all the network addresses discovered on the network segment.

- *Network layer matrix:* The network layer matrix captures the conversation between any two network addresses that have been discovered by the RMON probe. The MIB groups also capture the "TopN" count of the conversation pairs.

- *Application layer hosts:* This MIB group differs slightly from the network layer host group in that it records the protocol types used in a conversation pair. The protocol types are the ones listed in the protocol directory. For example, if two stations are conversing using HTTP, FTP, and telnet protocols, the application layer hosts MIB gets the information about these conversations based on the network layer hosts MIB and then starts recording the protocol counts based on protocol types HTTP, FTP, and telnet. These protocol types are listed in the protocol directory MIB.

- *Application layer matrix:* The application layer matrix captures the conversation between any two network addresses, by protocols. The conversation pairs are the ones listed in the network layer hosts while the protocols are listed in the protocol directory MIB.

- *User history:* The user history group collects historical data on a user based on the alarm thresholds programmed by the management stations.

- *Probe configuration:* The probe configuration group controls the various operating parameters of the RMON probe and is set by the RMON management station. The operating parameters include the probe capabilities and the system specifications like date and time.

- *Trap destination table:* The trap destination table configures the destination address for the traps generated by an RMON probe. The destination address is very likely to be the network management system.
- *Serial connection table:* The RMON probe can be configured to send the traps via out-of-band connection. This connection can be implemented as a SLIP connection using the serial port on the probe. The table stores all the relevant information needed to set up the SLIP connection.

The RMON-II specification also specifies extensions to the RMON-I MIB for RMON-II devices.

RMON and RMON-II compliant probes are available from Hewlett Packard, 3COM, SolCom, and Bay Networks (now Northern Networks). The hardware probe and the management software are priced from $5,000 to $10,000. Although most probes are interoperable with the management package from other vendors, it is better to go with one vendor solution. Some of the key points of consideration while selecting an RMON solution are listed in Tables 4.7 and 4.8.

4.4.2   RMON Product Offerings

The network management marketplace has some undisputed leaders.

Table 4.7
RMON Probe Evaluation Criteria

RMON Probe Features	
Number of RMON and RMON-II MIB groups supported by the probe	All or partial coverage
Interface types available	Ethernet, fast ethernet, FDDI, token ring
Is there an out-of-band telemetry port?	This could be a serial port or an ethernet port
Total number of ports available	This could range from one to six
Total memory available on board	This could range from 32 Mb to 128 Mb
Is there a command line interface for remote login?	Some probes can only be configured by the GUI interface
Cost	This could range from $2,500 to $8,000

Table 4.8
RMON Management Software Evaluation Criteria

RMON Management Software Features	
Number of RMON and RMON-II MIB groups supported	All or partial
Reporting capabilities of the management station. Is it Web based?	Some are Web based with each graph being presented as a GIF image
Level of integration with a network management system	The probes are represented as managed network devices and can send traps to the NMS
Ability to archive data and present a historical analysis	The archiving method could be simple flat files or an RDBMS system like Oracle
Runs on which operating system?	Sun Solaris, HP-UX, IBM AIX, or Microsoft NT
Cost	Ranges between $8,000 to $10,000. The final package will most likely be a collection of management and reporting tools

Hewlett Packard's HP OpenView NetMetrix Suite

HP OpenView NetMetrix is a family of products that provides the RMON management and reporting tools. The NetMetrix management and reporting tools work with the HP LAN probes and hardware-based RMON monitors. HP LAN probes implement all the MIB groups in RMON and RMON-II. The HP LAN probes are available for ethernet, fast ethernet, FDDI, token ring, and ATM networks. HP RMON probes are also available for WAN links supporting speeds up to DS3. The RMON reports are published as HTML documents for easy access via any standard browser. A sample report from HP NetMetrix reporter is shown in Figure 4.5. The figure shows the daily protocol distribution in a network segment.

3Com's Transcend Traffix Manager

Transcend Traffix Manager provides a powerful tool for network traffic analysis, trend analysis, and network troubleshooting using the data from standalone RMON/RMON-II probes as well as the embedded agents. The reports are available in a Web format. The RMON data can be stored in RDBMS for long-term historical analysis. The Traffix Manager provides standard reports for connection activity, specific device or group activity, TopN connections, devices, or network segments. Traffix Manager is available for Sun Solaris, HP UNIX, IBM UNIX, and NT platform.

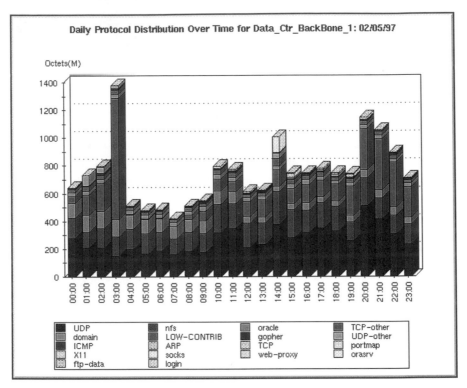

Figure 4.5 RMON traffic reports.

4.5 Service Management

Network management is a well-understood concept, while the service management concept is relatively new but fast becoming an important aspect of overall service delivery within an enterprise. When you deploy intranet or extranet networks with support for service management, you will be able to provide service level guarantees regarding various QOS parameters like latency, bandwidth availability, and uptime of the service.

A service-management framework includes end-user management, application availability, and data management including security and performance management. As you implement organizational intranets or extranets, ask yourself "Who are my customers?" Each group within the company needs to identify the service it is providing to its internal and external customers. The next step would be to identify the scope of service and then set SLAs based on the quantifiable baseline for the service. The final stage is the management of these SLAs and providing appropriate feedback and reporting mechanisms to monitor the conformance to SLAs. The conformance will lead to a high QOS, satisfied customers, and a distinct competitive advantage for the organization's success.

Service management encompasses the following aspects:

- Configuration management, asset management, and software distribution;
- Data backup, disaster recovery, high availability, load balancing, and security;
- Performance monitoring, diagnostics, throughput analysis, response time, and latency measurements.

There is no single software from any company that allows you to manage all the aspects of service management. In fact, some of the aspects are completely nontechnical and even intangible, such as expectation management and user satisfaction. You will have to mix and match software tools that allow you to manage one or more service management aspects. HP OpenView Suite has a comprehensive tool set dealing with the network, security, deployment, storage, and application management.

4.6 Conclusion

For a successful deployment of an intranet or an extranet service it is critical to have network and service management strategies in place before their deployment. An afterthought will result in a less optimal service and high support cost. A good SNMP and RMON management system will provide you with all the data needed to continuously monitor and enforce the SLAs with the internal and external customers.

5

Security Components of Intranets and Extranets

The key aspect of an intranet or an extranet implementation is the security of the information asset. By their very nature, intranets and extranets are closed to the general public and provide selective access to their user base, internal or external to the organization. The great wall protecting the intranet or extranet consists of many security components and covers a wide range of terrains. Their single purpose is to protect the information assets of the organization. For example, the assets may include the intranet applications providing various organizational services such as:

- Human resources for managing employee benefits, salary, and claims;
- Sales and marketing sites for competitive information;
- Information repositories for R&D and the engineering group;
- Software repositories for the developers;
- Internal e-mail and groupware services;
- Corporate financial services.

The extranet applications could be providing services like the following:

- Sharing customer information for order processing;
- Sharing confidential design and technical documents for a new product rollout;
- Streamlining the supply-chain processes in the production cycle.

As you know, these information assets are extremely critical to the smooth operation and the competitive well-being of a company. In today's competitive age, information is power, and companies want to give this power to their employees, vendors, and customers, empowering them with information so they can make sound business decisions based on good data. The other side of the coin is that the same data in the wrong hands, internal or external, could pose a serious threat to the company's competitive edge and financial well-being. A breach in security could result in tangible or intangible loss. The tangible loss could be of customers, technological edge, physical server, and network hardware, while the intangible loss could be of productivity, company reputation, credibility, and prestige. This is not an exhaustive list of the various reasons to keep the company's network secured from outsiders, but it helps highlight the seriousness of intranet and extranet security.

In this chapter we will explore these security components, their design considerations, and various tools available for their implementation.

5.1 Security Framework for Intranet and Extranet

The security framework is the first step toward the establishment of a secure intranet or extranet implementation. The security framework encompasses the following areas.

Security Policy Framework Consisting of Security Policies, Standards and Guidelines, or Best Practices   The security policy formally defines the rules by which all participants must abide. The framework also includes the security standards, guidelines, and best practices for accessing or distributing organizational information assets and resources. The security guidelines and best practices should include all aspects of security including physical security, network security, application-level security, and access to and from external networks.

Organizational Roles and Responsibilities   What roles do various individuals play in the organization and what is their responsibility to educate, implement, and abide by the doctrine of the security policy? For example, an employee in the role of a manger or a supervisor has a responsibility to make sure that all employees, contractors, vendors, or partners understand the organizational security policies and abide by them. The buck stops with the manager, if the security is breached due to noncompliance within his or her circle of influence.

Baseline Standards Should Exist for All Components of an IT Infrastructure Including:

- *Server security.* Server security includes the process of securing operating systems (OS), their file structures, administrative accounts, user accounts, and server processes. Examples of various common OS platforms are UNIX, NT, MAC, and mainframe. Each OS platform needs its own care and feeding. The OS vendors issue security or performance-related patches from time to time. The system administrators and the network administrators need to be aware of the latest patches and the holes they fix. It is also a good idea to create a formal checklist of standard installation procedures for each OS. For example, NT creates a "guest" account with no password by default. Another example would be all standard services are turned on by default in a Sun Solaris installation. If the installation across each main platform is standardized, it is easier for the administrators as well as the help desk personnel to troubleshoot the problem.

- *Network security:*

  - *Network devices like routers and switches.* The baseline standards for network devices should include standard configuration and naming conventions. Access to the routers must be thorough RADIUS or TACACS. This type of access mechanism leaves a nice audit trail as to who logged into a router and when. The router logs should be sent to a "syslog" server. The log files on the syslog server can be parsed for specific message strings or a pattern of messages. This type of analysis can provide valuable data for alerting administrators on a variety of issues, from network attacks to performance-related proactive alerts. Additional security standards on the storage of passwords on the network devices, specifying who has administrative access from which network or location, will strengthen the security of the network operation.

  - *Network management platforms.* The network management standard on SNMP access to servers or network devices is critical in securing the network operation. Since SNMPv1 and v2 implementations pass the SNMP passwords in clear over the network, the private and public community strings should be kept secret. The write operation to remote devices should be logged and monitored. Different groups within the company may be configured with read only or read/write access depending on the need.

  - *Network level filtering using routers or firewalls.* One of the first lines of defense in an intranet or an extranet implementation is the ACLs or rules on the routers and firewalls on the periphery of the network. At the router level you can control packet-level access to the network or an application like the Web from a specific source. For example, ACL 134 in the sample configuration given below allows access only to Web services on port 80 from 204.70.134.0 subnet for any packet traveling outbound on the ethernet interface number 3.

```
interface Ethernet3
description Remote Access subnet
ip address 207.80.127.129 255.255.255.192
ip broadcast-address 207.80.127.191
ip access-group 134 out
access-list 134 permit tcp 204.70.134.0 0.0.0.255 any eq www
access-list 134 deny any any
```

Firewalls provide an additional level of granularity in providing access to the control and log data of each access. In the previous example, you can program the router to provide access control at the Web (port 80) level, while in the case of firewalls, you can control access at the Web URL (uniform resource locator) level. You can additionally specify the time of access and the level of logging for audit purposes. This level of control is great when you are providing extranet access to your vendors, customers, and partners. Baseline standards on the functionality of firewalls, the access control rules, and the audit log requirements are essential for secure operation of an intranet or an extranet.

- *Application security including security of:*
  - *Web applications, including CGIs and server side includes.* Baseline standards for using CGI or server side includes on intranet or extranet Web servers will help in avoiding some of the vulnerabilities of such programs on a Web server. Additional standards for function libraries used in the development of Web-based application programs should also be defined as good security measures.
  - *Data backup and recovery.* Data backup and recovery standards are absolutely crucial for sound operation of an intranet or extranet. The greatest information asset the security standards are trying to protect is the data. The data could be in the application servers, application databases, or on a network drive. If the data is corrupted due to the failure of a system or its drive or due to a security incident, the organization will have to rely on the backup data to recover quickly from any such incident. Baseline standards on the frequency of the backup and the level of backup are an essential part of uniform pro-

cedures throughout the enterprise. An example of a backup standard could be a weekly full backup (level 0) of all systems with daily incremental backup. The backup of various systems could be staggered so that it does not put an inordinate amount of load on the network. If there is a large amount of data backup per LAN, you might want to consider using a separate network for backup. An off-site storage facility for the backup tapes is also a good idea.

- *Configuration management including version control.* Various intranet and extranet systems, network devices, and applications are subject to frequent changes in response to changing business environments. A baseline standard for version control and configuration management is essential in maintaining the integrity of the device configurations. The configurations of routers, their access control lists, server applications, server configurations, account data, and application data should be maintained in a secure environment. If the current configuration of the devices is corrupted due to a system failure or a security incident, you can quickly recover the lost configuration from the configuration management system and can even go back a version to reload an untainted version.

Risk Management Including Risk Assessment, Risk Quantification, and Risk Prevention   The process of risk assessment includes identifying various sources of threat and their motivating factors, how likely it is that these threat sources can gain access to your information assets, and finally, what the cost of protection from such threat is. Organizations need to spend a lot of time upfront in conducting a risk assessment. The time will be well spent!

Incident Response and Escalation Procedures   The Federal Computer Incident Response Team's (FCIRT) definition of the term *incident* defines it as "an adverse event in an information system and/or network or the threat of an occurrence of such an event." An event is defined as "any observable occurrence in a system and/or network." The term *incident* covers events like malicious code attack, unauthorized or unintended access, unauthorized or unintended utilization of service, disruption of service, misuse of the service, espionage, or hoaxes. Any response to these incidents needs organizational resources. (FCIRT's Web site is http://www.fedcirc.gov/.)

In order to best utilize the limited resources within an organization, decide on incident response procedures prior to the incidents. A coordinated and well-planned response will increase the effectiveness of the response as well as reduce the cost of the response. If the response is not coordinated, it will result in potential chaos, disruption of service, and loss of important data that could be used in identifying the incident source.

When an adverse event is detected there are two types of responses:

1. Immediately contain the event to the smallest number of systems and/or networks.
2. Watch and pursue the perpetrator while accepting the risk to the systems and/or networks.

The type of response depends on the type of incident and the type of service affected by the event. In either case, the escalation process needs to be in place to send the information about the event to upper management and customer service representatives.

All organizations, depending on their size, are encouraged to set up incident response teams and an incident response plan. All the administrators within the organization can then coordinate their responses through this central team. The central team can also bring in outside entities like the Computer Emergency Response Team (CERT) or law enforcement agencies like the FBI to identify and prosecute the perpetrators. CERT also runs a training program for training your own incident response teams and discussing how they can work with CERT. There are localized versions of CERT, performing similar functions on each continent. (CERT's Web site is http://www.cert.org/.) The incident response plan needs to have following base content:

1. Goals and objectives of the plan;
2. Process for identifying the incident and its scope;
3. Clear process for incident notification and management escalation (for example, the first group of managers notified) and customer notifications. Also, clear instructions for means of notification such as phone, pager, or e-mail. If the primary contact for each category does not respond, who is the backup for the primary?
4. Directives for handling an incident should include the following information:

   • When to use the capture and prosecute strategy;
   • When to use a containment strategy;
   • How to coordinate with internal and external incident response teams like CERT;

- FCIRT and vendor support teams;
- How to capture the evidence of intrusion;
- Steps for restoring the affected data or any other assets like the OS or a device configuration to recover the affected systems;
- How to file an incident report including the report format.

5. Documentation and analysis of the incident including:

- Recording of exact sequence of events;
- Customer impact and number of customers affected;
- Cost analysis of the incident, including loss of hardware, software, data, and loss in productivity;
- Steps taken to prevent an occurrence of such an event in the future (this could include changes in the operational architecture, changes in configuration of hardware or software, changes in policy, or changes in processes);
- Adding the documentation to internal knowledgebase for future reference.

A sample incident report format is shown in Figure 5.1.

6. Process for handling the customer queries and queries from the media. Process for handling any public relations (PR) interviews and announcements.

User Training   User training is extremely important in raising the awareness of security, making sure that each individual understands his or her role and responsibilities, and the use of the security tools like antivirus software, disabling a Java execution environment, setting right permissions during the system installation, or something as simple as making sure the user hard drives are not shared to the world. In most organizations, security is considered a necessary evil, instead of an essential part of a successful enterprise. It is the management team's job to change this negative perception of security and get everyone in the organization involved in the corporate security program. You know the security awareness program is working if you are stopped by a janitor in the off-office hours and asked to show your access authorization in restricted parts of the building.

Continuous Monitoring and Feedback Mechanism   See Figure 5.2.

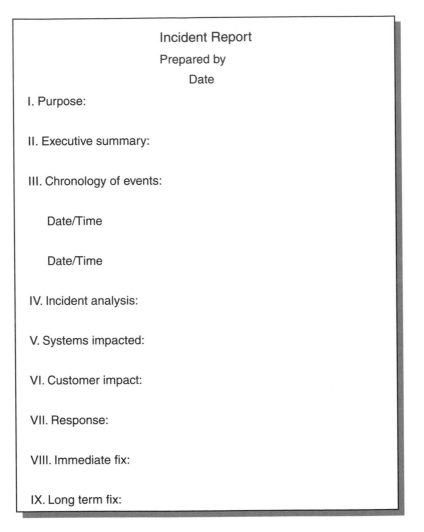

Figure 5.1 Security incident report.

In today's fast-paced world, maintaining a secure environment is a continuous process. The needs of the organization are in constant flux. As new tools and software are deployed, there is potential for new threats. The security group has to go through a continuous process that involves:

- Risk assessment;
- Implementation of new tools, patches, and regular training programs;
- Enforcement of the new policies;
- Regular audits for policy compliance.

Figure 5.2 Risk management feedback loop.

5.2 Developing a Security Plan

When you are developing a cost-effective security plan for your environment, it is good to ask the following questions:

- *What are we trying to protect?*
  Identify the tangible or intangible organizational assets that you are trying to protect. The tangible assets could be the data, hardware, software, people, or documentation, while the intangible asset could be the trust of your customers or the company's reputation with its vendors.

- *From whom are we trying to protect ourselves?*
  Internal users and/or external threats. The threats themselves could be based on a systematic or random attack.

- *What is the likelihood of these threats?*
  Certain types of threats are more apparent than others. This exercise will allow you to prioritize the list of potential threats and develop policies to combat these threats. For example, it has been proven that the most significant threats to the information assets are from within the company and not from outside the company. The policy measure to combat this top priority threat is to provide access to information on a need-to-know basis. The next obvious question would follow:

- *How would we go about implementing such a security policy in the most cost-effective manner?*
  It is very important to weigh the cost of protecting the information asset versus the cost of recovering from the threat. The cost of protecting

from a threat should always be less than the last two factors. If you are trying to protect from internal threats, the best security plan is a plan of deterrence.

- *Is the policy enforceable?*
  Some policies may be so cumbersome that they break the whole security framework. For example, if you make a policy to "age" the passwords every 15 days and have strict password selection guidelines, the users are going to start hating the policy and will start posting the passwords on the monitors or a piece of paper on their desks. This breaks the whole security framework that you are trying to implement.

- *What can be done to improve the security?*
  Development of a cost-effective security plan is a continuous process. The security policy document is a living document that needs to adapt to the changing threat environment.

- *What are the tradeoffs?*
  My favorite security oxymorons are, "simple security," "security service," and "cheap security." You will always find yourself in a situation where you have to give up something to get something. The more security measures you have, the more you will have to give up on ease-of-use, services offered, or just plain money. As security implementations get complex, to accommodate strict policy measures, the cost of maintenance goes up as well.

5.3 Security Tools

The tools available for implementing the security policies based on the security framework can be categorized as follows (see Figure 5.3):

- Prevention;
- Detection;
- Correction.

Figure 5.3 shows the intersection of each category with the three main areas of security.

5.3.1 Prevention

The prevention category focuses on the various tools and strategies for preventing an occurrence of a security breach at the network, server, and applica-

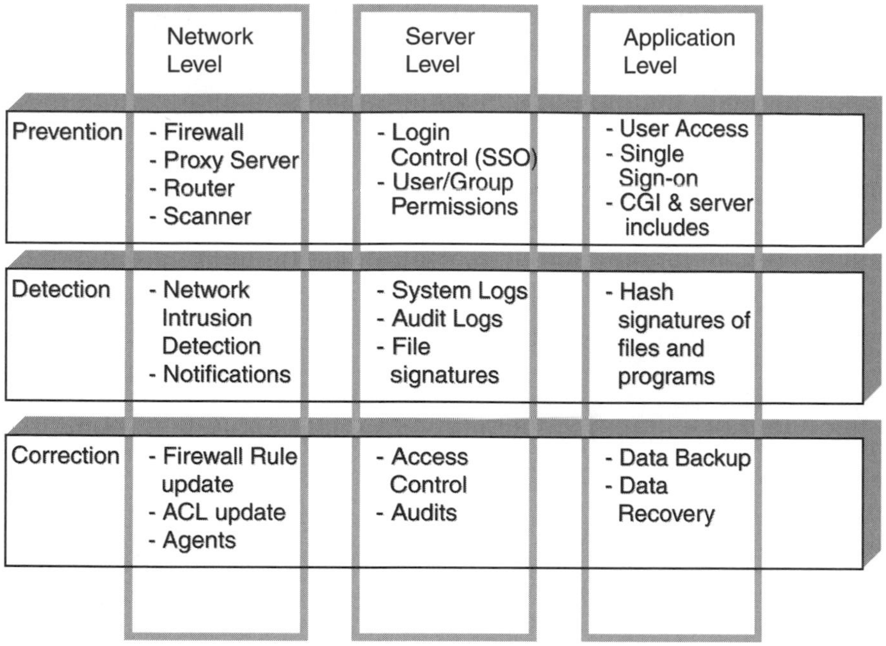

	Network Level	Server Level	Application Level
Prevention	- Firewall - Proxy Server - Router - Scanner	- Login Control (SSO) - User/Group Permissions	- User Access - Single Sign-on - CGI & server includes
Detection	- Network Intrusion Detection - Notifications	- System Logs - Audit Logs - File signatures	- Hash signatures of files and programs
Correction	- Firewall Rule update - ACL update - Agents	- Access Control - Audits	- Data Backup - Data Recovery

Figure 5.3 Security tools for network, server, and application security.

tion levels. For example, at the network level firewalls are likely to be the first and last line of defense for network-level security in most organizations. The firewalls could be application firewalls, routers with ACLs, or application-level proxies. We will take a close look at various types of firewalls and their implementations later in this chapter. Examples of firewalls are Checkpoint's software-based Firewall-1 solution, RADGARD's hardware-based firewall, or Aventail's application-level proxy firewall based on SOCKS-5. Another set of tools for incident prevention consists of network scanners. The network scanners probe the outer and inner defenses of the operational environment for well-known vulnerabilities.

Network-based scanners can also probe the vulnerabilities at the server and application level. The vulnerabilities could be OS related, configuration related, or software related (for example, configuration of user and group privileges for a network drive or OS-related patches). Examples of network scanners are Internet Security System's (ISS) Internet Scanner, Cisco System's NetSonar, Secure Network's Ballista Security Auditing System, or publicly available SATAN.

Application- and system-level access control is normally a second line of defense. The tight control over the access to the applications in a secure manner within an intranet or over the extranet is very important. One way to accomplish this goal is via a single sign-on solution. A single sign-on solution allows you to strengthen the security of your infrastructure by eliminating internal trust relationships but still allowing the user to log into various applications with a single log-in. The user can continue to maintain a separate user ID and password for each application or a system, but does not have to remember all the user ID and password combinations. We will look at the pros and cons of various single sign-on solutions in the market later in this chapter.

5.3.2   Detection

The detection category focuses on the tools and strategies for detecting the breach in security as soon as it occurs. Intrusion detection tools are available in the market that can detect a network-, system-, or application-level intrusion or an attack based on promiscuous monitoring of the network or a system using distributed agents. The agents could be software agents or hardware probes. What are these agents and intrusion detection tools looking for? The various activities of interest for such tools are listed below (see Table 5.1).

5.3.3   Correction

The correction category focuses on the tools and strategies for rapidly responding to the breach in security as well as a continuous security-improvement process. The correction tools provide functions like automated firewall update and reconfiguration of remote agents. One of the important strategies for data correction and timely response is to have a good data backup and recovery strategy. The operators and administrators must be well trained in backing up the good data as well as recovering the good data from the previous backups. The tracking of an attack in progress is useful if you plan to prosecute the cracker and have the trained resources to do so (provided, of course, that you are alerted as soon as the attack begins!).

5.4   Data Security

In the case of an intranet, although the content technically does not leave the corporate network, it is still vulnerable to various types of attacks from malicious users inside or outside the organization. The users could be employees, on-premise contractors, or vendor employees. Security holes in the intranet implementation also make it possible to mount an external attack on the intranet. This problem becomes more acute in the case of an extranet, where there are

Table 5.1
Security Incident Detection Tools and Their Detection Criteria

Activity Logs	Unusual Activity	Detection Tool
UNIX Syslog	- Multiple failed login attempts - Various user ID's being tried in succession at odd times of a day - Number of failed attempts to gain root privileges - Attempt to access/change system critical files or programs - Attempt to change user/group permissions on different accounts - Missing files - Port scanning attempt on various well-known ports	System Agent Ex. Tripwire TCP-Wrapper
Router Syslog	- A large burst of activity or a pattern of activity from a certain source or range of addresses - Failed multiple login attempts - Failed attempts to enter "enable" mode	Log Parser
Router Syslog	- Failed login attempts - Unusual panic condition - Port scanning attempts on the router	
Network Monitor log	- Monitors TCP/UDP type of packets and the packet patterns of a known attack type	- Netman - ISS
Application logs	- Large number of error log entries - Unusual access attempts	- Log Parser
NT logs - Audit Logs - Event Logs	- Log of invalid file access, application related logs, security logs, and the system logs	- Event Viewer

one-to-many trust relationships with external entities. In all these scenarios it is essential to maintain security of the information asset. The information assets can be protected if the security architecture based on the security framework accomplishes the following goals (see Figure 5.4):

- Data confidentiality;
- Data integrity;
- Good access control features with solid authentication;
- An ability to conduct extensive security audits.

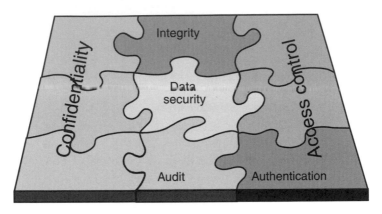

Figure 5.4 Goals of a security framework.

In this section we will look at various means of achieving these goals.

5.4.1 Data Confidentiality

There are various means of maintaining data confidentiality. One way would be to prevent any access to the data in the first place. This part will be covered in Section 5.4.3. Another means of achieving data confidentiality is by encrypting the data using various cryptographic techniques. This provides data privacy on the system as well as during data transmission. In this section we will concentrate on the major cryptographic algorithms and their implementation in commercial products that could be used to provide data confidentially by an intranet or an extranet implementation.

Introduction to Encryption

Put simply, encryption is a method of transforming readable data into a non-readable format which can be recovered using a special key. A simple analogy would be locking a document in a safe (see Figure 5.5). As long as the document is in a safe, nobody can read it. You can now transfer the safe from one location to another location. If you have the valid key for the safe-lock, you can retrieve the document from the safe and read it.

What is really going on behind the scenes?

The encryption algorithm processes the incoming plain text data with the encryption key into a cipher text data (see Figure 5.5), an encrypted version of the original data. The encrypted data is decrypted using the same or a different key, depending on the encryption algorithm used. When the data is in an encrypted format, the data can be stored or transferred within a network without fear of being read. The key component of this security mechanism is of course the management of keys and how they are stored securely. If the encryption key is compromised or the system storing the plain text data is compromised, encryption will not help you. Thus, encryption is one part of the whole data security puzzle and hence other pieces of the puzzle are equally important.

Some of the encryption algorithms can also be used to provide authentication by offering digital signatures.

There are two primary encryption techniques:

1. Secret key encryption, also known as the symmetric key encryption;
2. Public key encryption, also known as the asymmetric key encryption.

Symmetric Encryption

In the case of symmetric encryption, the same key is used to encrypt and decrypt the data, as shown in the "safe" example in Figure 5.5. The challenge in symmetric key encryption technique is the secure exchange of the encryption key (K_{en}) between two communicating parties. This is especially true if the two parties are not in the same location. Examples of symmetric key encryption algorithms are shown in Table 5.2.

Table 5.2
Symmetric Key Encryption Algorithms

Symmetric Key Algorithms	Comments	Key Length
UNIX "Crypt"	- Not secure	Variable
DES	- Data encryption standard	56 bit
	- Not so secure anymore	
	- Use the 3DES version	
3DES	- Triple DES, highly secure	56 bit/key
RC2	- Rivest's cipher (RC)	Variable
	- Block cipher	
RC4	- Stream cipher	Variable
RC5	- Block cipher	User defined
IDEA	- International data encryption algorithm	128 bit
	- Used by PGP	

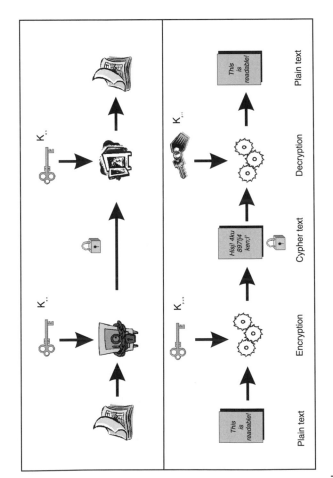

Figure 5.5 Encryption schemes.

Asymmetric Key Encryption (or Public Key Cryptography)

Asymmetric key encryption solves the fundamental problem of the secret key encryption, the secure exchange of the encryption key. In the case of asymmetric key encryption, two separate keys are used to encrypt and decrypt the data, as shown in the second example in Figure 5.5. Each communicating side has a key pair consisting of a private and a public key. The public key is made available to the public. The sender uses the receiver's public key (K_{pub}) to encrypt the message, while the receiver uses the associated private key (K_{pri}) to decrypt the message. It is absolutely essential that the private key is truly private. Examples of asymmetric key encryption algorithms are shown in Table 5.3.

In order to understand the workings of an asymmetric key encryption algorithm, we have selected the most popular public key algorithm in use, the RSA algorithm.

RSA Encryption

The RSA cryptography system provides one of the most well-known implementations of a public key cryptography system. The RSA system is based on an algorithm developed by MIT professors Ronald Rivest, Adi Shamir, and Leonard Adleman (RSA) in 1978. The RSA cryptography system can be used for encryption and authentication (digital signatures). We will explain the working of the RSA system based on Figure 5.6.

Table 5.3
Asymmetric Key Encryption Algorithms

Asymmetric Key Algorithms	Comments	Key Length
Diffie-Hellman	- Provides a way to exchange keys	Variable
	- Digital signatures are used to exchange the keys with authentication. This prevents man-in-the-middle-attack.	
	- Highly secure	
RSA	- Used for both encryption and digital signatures	Variable
	- Highly secure	
DSA	- Digital signature algorithm	Variable (512–1,024 bits)
	- Used for digital signatures only	
	- Developed by NSA	
	- Highly secure	

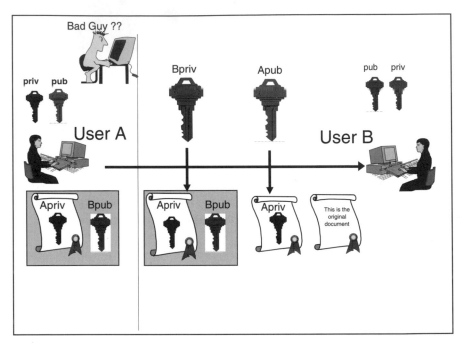

Figure 5.6 Operation of RSA encryption scheme.

When User A wants to communicate securely with User B, it requests User B's public key. User A can also get User B's public key from a public key ring maintained by MIT (http://pgpkeys.mit.edu:11371) or the PGP (pretty good privacy) corporate key server. Once User A has B's public key, B_{pub}, the key is used to encrypt the message being sent from A to B. When User B receives the message, User B uses its private key (B_{pri}) to decrypt the message. If the bad guy captured the encrypted message he or she cannot decrypt the message since he or she does not have User B's private key. The message can also be digitally signed by User A. We will take a look at digital signatures later in this chapter.

5.4.2 Data Integrity

Data integrity provides an assurance that the data has not been modified in any way. The data integrity is used in UNIX systems to make sure that none of the system's critical files have been tampered with, while PGP uses data integrity to maintain the integrity of e-mail. System and network surveillance tools use the data integrity test to detect for any signs of intrusion. There are two ways to establish data integrity:

- Cryptographic checksum;
- Cryptographic hashing.

Cryptographic checksum is normally used for detecting any changes to the data during a file transfer or to conduct a quick check on the correct file. This is not a strong method of checking data integrity since it does not detect multiple changes in the bitstream that may still result in the original checksum. This behavior makes it quite useless for detecting any security breaches. An intruder could very easily substitute original programs with Trojan horse programs that have an identical checksum. A more secure method of establishing a baseline for data integrity is required. Cryptographic hashing provides one such method.

Cryptographic hashing uses a mathematical hash function to transform any variable length input into a hash value of fixed size. If "x" is the input and H{} is the hash function, then

operation H{x} = y (Fixed size, for example 128 bit)

produces "y," a fixed width output, which is much smaller in size to the large input. Some of the characteristics of the hash function are:

- It is one-way hash, (i.e., it is very difficult to reverse the process and derive the input from the output);
- It is deterministic (i.e., given the same input, it produces the same output);
- For a unique input, it produces a unique output.

Cryptographic hashing is also used for creating digital signatures. Some of the prominent examples of hashing algorithms are MD2, MD4, MD5 (message digest 5) and SHA (secure hash algorithm). All message digest algorithms produce fixed output of 128 bits, while the SHA algorithm produces 160-bit output. The MD5 signatures of mission-critical data and configuration files are handy in detecting any intrusion. If the intranet application server is a UNIX server, programs like Tripwire accomplish exactly this task with a high degree of efficiency.

Note on Tripwire   Gene Kim and Gene Spafford wrote the Tripwire program at Purdue University. Tripwire creates an MD5 checksum of the files and directories that have been configured for monitoring by the administrator. The checksum provides a signature of the file. If the file is changed in any way, the new signature generated by the Tripwire program will not match the signature before the modification, thus providing an accurate means of detection. This information is then reported to the administrator. The Tripwire program can be downloaded from ftp://coast.cs.purdue.edu/pub/COAST/Tripwire.

5.4.3   Data Access Control and Authentication

Up till now we have covered the data confidentiality and data integrity aspects. These two aspects ensure that we have good data. Now it is important to control the access to this good data. The best access control mechanism includes:

- Physical access control;
- Network and system access control.

Physical Access Control

The physical access control includes access to:

- *Premises.* Controlling who gets access to the physical premises by way of badges and visitor passes is important. The visitor ID should be prominently displayed so that everyone knows the status of the individual as a visitor. Employees should be educated on the process of handling visitors. For example, the site-security policy might state that visitors are not to be left alone; they need to be escorted by an employee right from the time they enter the premises to the time they exit the premises. Large signboards with "Authorized Personnel Only" need to be displayed at secure locations. All alternative routes of accessing the premises should be secured. There is no point in having a strong front door when all your side doors are wide open.

- *Network.* In the case of an intranet LAN or a WAN connection, you need to control access from publicly available network access points like the conference room or visitor cubes. The network connection at these places needs to be isolated from the internal network. There is always a danger that someone could add a physical probe to your network that captures traffic for a couple of days which can then be analyzed for captured passwords.

- *Workstations and servers.* The physical access to the servers or workstation also needs to be restricted to authorized personnel only. Make sure the keys to the server cabinets in a data center or docking stations on the desktop are secured and not easily available to a passing individual. The power connection to the server hardware needs to be secured from intentional or unintentional disconnect. Removable hard drives in desktops and laptops need to be secured so that they are not easily removed.

Note It has been my observation that a significant amount of knowledge and data is lost when employees or contractors are leaving a project or the company. Their computers may or may not be collected. If they are collected, the individuals might erase the hard drive before handing over the computer. To avoid this type of loss the security policy should dictate that the managers in charge of departing employees or contractors need to instruct the individuals not to erase the data and allow deletion of personal data under supervision. This problem becomes very acute if the employee or the contractor is leaving under unfavorable circumstances.

- *Backup tapes and drives.* Data backup is an important aspect of data security. In most IT enterprises, the data is backed up on multidrive jukeboxes or a central shared drive. Physical access to the backup tapes or the drives is critical in maintaining data security. Today, a single backup tape can hold a multigigabyte worth of data. Local backup solutions like ZIP® drives and ditto® drives can store PC data from 100 MB to multigigabytes. It is very easy for someone to pocket these backup media and walk away with valuable data. Many applications store the user ID and associated password information in the local machine. A backup tape containing these files is a treasure chest for a malicious user.

  How do you battle against such a threat? The first and foremost solution is to secure the data in locked cabinets with added access control based on the sensitivity of the data. Second, you can use more exotic solutions that will encrypt the data on the hard drive and hence the data on the backup tape is also encrypted. These tools encrypt and decrypt the data on the fly as it is requested by the user or by an application. Even applications like Oracle can store data in an encrypted format. A backup of this encrypted data is protected from any hacking attempts.

- *Printouts and discarded papers.* Even though we are moving towards a paperless society, we still print a large number of our data. How many times have you gone to a shared network printer to collect your printout and found it missing? You probably assumed that it never printed or was taken by mistake by another user. How do you know that someone did not take it intentionally? I am not asking you to be totally paranoid, but it is better to be paranoid than to lose important data to industrial espionage. Even discarded paper with remotely important or confidential data should be sent to a shredder. (How do you know the person shredding the papers is reliable!)

Network and System Access Control

The network-level access to the data can be controlled by a policy-based configuration of a firewall ruleset or a router access control list. If you are connecting your vendors, customers, or partners to your intranet via an extranet network, you could dedicate a subnet within a class C address space or an entire class C to the extranet network and servers within that network. LAN connections within conference rooms and visitor cubes should be isolated at the network and physical levels. Again, the address space in this subnet may be isolated from routing internally by clever use of IGP routing and use of access control lists.

The system-level access control is achieved by controlling the system accounts and the trust relationships within or outside the enterprise. An account request on a system needs to be properly authenticated and documented. Even the level of privileges associated with an account should be on a need-to-have basis. By the same token, the login IDs of the leaving personnel need to be deactivated immediately and removed only after complete backup of the data has been completed. Centralized security administration solutions like single sign-on or token authentication solutions like SecurID are excellent in implementing and enforcing such policies. We will take a look at both solutions later in this chapter.

To simplify a user password management process, administrators resort to setting trust relationships between multiple applications or domains. This is quite a disturbing trend since this method lacks any audit control process for tracking access behavior of an individual across multiple domains or applications. This trust relationship also leads to an all-or-nothing approach of security. Consider a situation where a contractor is granted an account on an NT domain that has a trust relationship with six other domains. Unless there is a more granular control within the trust rules at the user account level, the contractor now has access to a large amount of data, thus increasing the security risk. In the case of a UNIX account setup, similar risks exist if the account has a trust relationship with accounts on other systems.

5.4.4   Authentication

The strong access-control mechanisms need to be complemented with strong authentication schemes. Different authentication schemes provide different levels of security. As in any security mechanism, there is naturally a tradeoff between the level of security versus the complexity versus the cost. We will take a look at some of the prominent authentication mechanisms that could be deployed in an intranet or an extranet and explore their pros and cons.

1. Password-based authentication;
2. Single-step/two-step token authentication;
3. Kerberos;
4. Digital signatures;
5. Use of certificates;
6. Single sign-on;
7. Dial-up PAP/CHAP authentication for remote access.

Password-Based Authentication

The password-based authentication assumes that you have an account with a unique user ID to get access to data on an intranet system. The password mechanism is the most popular mechanism in the industry and it is the least secure. There is no inherent lack of security in the mechanism itself but it is the application of the mechanism that causes security breaches. How many times have you seen a password being written on a note pad or a notebook or as a plain text file in someone's computer? Individuals resort to these methods because the passwords are too complex to remember and need to be changed frequently. A typical security policy might dictate that the password should be changed once a month and should be at least eight characters long with a mix of alphanumeric and nonalphanumeric characters. This leads to the "Post it!" problem.

At the other end of the password spectrum, if the passwords are simple and easy to remember, they can be cracked very easily with a simple dictionary or a social engineering attack. The best password scheme should strike a balance between the degree of complexity in a password and frequency of its change. There are quite a few programs on the Internet that can generate passwords that strike such a balance. Consider the following examples: r0ck2r0ll replaces *o* with *0*, r1ck?t1ck replaces *i* with *1* and adds a *?* Administrators can suggest these programs to the users. Administrators can also run periodic password cracking programs like, guess what, "Crack," or a crack variant, to check on the strength of the passwords. You can find some of the password cracking programs at the following sites:

NT password cracking programs:

- http://www.l0pht.com/
- http://www.somarsoft.com/ntcrack.htm

UNIX password cracking programs:

- http://ciac.llnl.gov/ciac/SecurityTools.html

One-Time Passwords (OTP)

In most password-based authentication systems, the password is likely to be transmitted in clear over the LAN or WAN link. There are many sniffer programs, like Snoop on UNIX or NetXRAy for PCs that can capture the network traffic and the password exchange between a client and a server. Since the passwords are sent in clear, the sniffer operator now has access to the account information like the user ID and the associated password to be used later. Is there any way to prevent this type of security compromise, also known as the "replay attack"?

There are two methods by which this type of security compromise can be prevented:

1. Use one-time passwords;
2. Encrypt the password during transmission.

In this section we will deal with password schemes that deal with one-time passwords, while the next section deals with the password encryption schemes.

The OTP concept was first introduced by Leslie Lamport in "Communications of the ACM" in November 1981. Neil Haller from Bellcore introduced the first implementation, known as S/Key™ software (S/Key is a trademark of Bellcore). S/Key was already in use within Bellcore for two years before it was introduced to the world. The Bellcore paper can be found at following site:

ftp://ftp.bellcore.com/pub/nmh/docs/ISOC.symp.ps

The software distribution can be found at the anonymous ftp site of Bellcore at the following URL:

ftp://ftp.bellcore.com/pub/nmh/

The work on OTP scheme is now being carried out by an IETF working group. The group's IETF charter can be found at the following URL:

http://www.ietf.org/html.charters/otp-charter.html

The group has published three RFCs:

1. RFC 2243: OTP Extended Responses;
2. RFC 2289: A One-Time Password System;
3. RFC 2444: The One-Time-Password SASL Mechanism.

How Does OTP Operate?

The OTP operation is illustrated in Figure 5.7.

Step 1: The server issues a challenge to the client. This challenge is sent in the clear. The challenge consists of all the values needed by the OTP generator at the client end to create a one-time password. These values consist of a random seed and the expected sequence number of the one-time password.

Step 2: The client OTP generator inputs the random seed received from the server and the client "passphrase" into a hash function H{}. The output of the hash operation is fed back into the hash function for *N*-iterations corresponding to the received sequence number. The hash function could be MD4, MD5, or SHA. The output of the hash function is a 64-bit value.

Step 3: The OTP is then sent to the server. At this point, even if a malicious user has both the password and the seed value from the earlier capture of the challenge, he or she cannot derive the value of passphrase. This is due to the one-way nature of the hash function. The password itself is only used once.

Step 4: The server passes the OTP through its own hash function and compares the value to the stored OTP value from a previous operation. If these values match, the client is authenticated.

Pros:

- The client passphrase never passes in the clear over the network.
- The password being sent from client to server is a one-time password. This protects the client-server communication from snoop or "replay" attack.
- No client password information is stored on the server, making the server invulnerable to a password attack.
- The algorithm itself is public, hence can be implemented by anyone.
- If you do not have an OPT generator, you can print a list of successive one-time passwords and carry it to an off-site location and use one at a time.

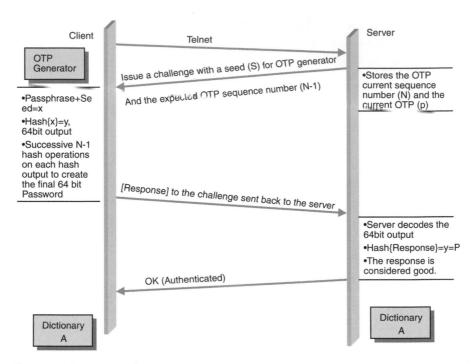

Figure 5.7 Operation of OTP scheme.

Cons:

- Some modification to the application on the server end and/or the client end is required.

One-Time Password With Token Authentication

All token authentication schemes involve two-factor authentication. The two factors being a PIN number or a passphrase and a number displayed on the token. Examples of token cards are SecurID card from Security Dynamics and Cryptocard from Cryptocard Corp. (see Figure 5.8). Let's first take a look at the workings of a SecurID card.

The SecurID card from Security Dynamics displays a unique number combination every 60 seconds. This number in combination with a numeric PIN number forms a unique password. This results in a unique password every 60 seconds. If a password is "sniffed" by a malicious user he or she will not be able to reuse that password. The SecurID card is shown in Figure 5.8. The PIN number protects against the loss of the physical card. Just like your bank ATM card, without the unique PIN number, it is almost impossible to use the card. The time bar on the card shows the countdown to 60 seconds. At the begin-

SecurID   CryptoCard

Figure 5.8 Token cards.

ning of each 60-second interval a full bar is shown on the left display of the card. The bar is depleted with the 60-second countdown. The use of a SecurID card requires use of an ACE server and an ACE agent from Security Dynamics. The ACE server provides a central database to manage all the accounts and the seed for the password synchronization on the server and the card. The ACE client is installed on target hosts and servers. The ACE client provides appropriate library functions to integrate with standard application programs. ACE clients are available for a wide variety of UNIX and NT platforms. SecurID is an example of single-step authentication. You can find more details on the Security Dynamics products at http://www.securitydynamics.com/.

Another type of token authentication is based on two-step authentication. The two-step authentication involves a challenge-response mechanism. One good example of a token-based two-step authentication scheme is CryptoCard's RB-1 token (see Figure 5.8). The RB-1 token has a unique PIN number (or a password) for each user. The CryptoCard provides a two-factor two-step authentication scheme.

The operation of a CryptoCard can be described in following steps (see Figure 5.9).

- The security group within an organization assigns a CryptoCard RB-1 with a unique DES key and a PIN number.
- The user initializes the CryptoCard with his or her PIN number. At this point the user is ready to log into a remote server. The operational steps illustrated in the figure are explained below:

1. The user connects to a network accessible service and receives a login prompt. User enters his or her user ID.

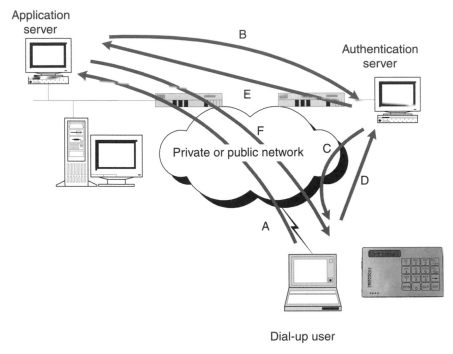

Figure 5.9 Authentication using CryptoCard.

2. The user's login request with the user ID is then forwarded to the authentication server.

3. The authentication server generates a random "challenge" to the user. The user inputs this challenge into the CryptoCard using the KeyPAD. (Remember that at this point the user has already initialized the card with his or her PIN number or a passphrase.)

4. The CryptoCard encrypts the challenge with its DES key and displays the results as the "response." User sends this response to the authentication server.

5. The authentication server computes the response using its copy of the user's key. If the result matches with the received response, the authentication is considered successful and an OK is sent to the application server. The application server provides access to the user.

You can find more details on CryptoCard at http://www.cryptocard.com/.

Pros:

- Secure one-time password mechanism;
- Universal access for a user of the card.

Cons:

- The cards tend to be fragile and could be lost or run out of battery power;
- The cost of the card;
- If you forget your card, you have no other recourse, you are simply locked out.

Kerberos

Another method that avoids the use of passwords on a clear channel is based on Kerberos. Kerberos was designed at MIT for an internal project called Athena. Kerberos has evolved from its first version to the latest version 5 (Release 1.0 Patch Level 4) and is documented in RFC 1510. Kerberos does not use public key cryptography. The Kerberos-based authentication system consists of:

- A client or a user workstation;
- A server providing the service;
- An authentication server;
- A ticket-granting server.

The goal of Kerberos is to provide a user access to any of the network-based services in a secure manner without compromising the user password. Kerberos achieves this by encrypting the user authentication data using DES. The whole concept is based on a central server (KDC) issuing a credential (ticket) to a user requesting a service from a network-based server so that he can present that ticket to the server and get access to the service.

A simplified explanation of Kerberos operation is shown in Figure 5.10.

1. User logs into his or her workstation with his or her login ID and requests access to a service on the service-providing server. The Kerberos client on the user workstation generates a request to the authentication server (AuthS) for credentials that will allow the user to request the necessary "service requesting credentials" from the ticket-granting server (TGS). The ticket-granting server ultimately issues the required credentials (or tickets) to the user for various services on the network.

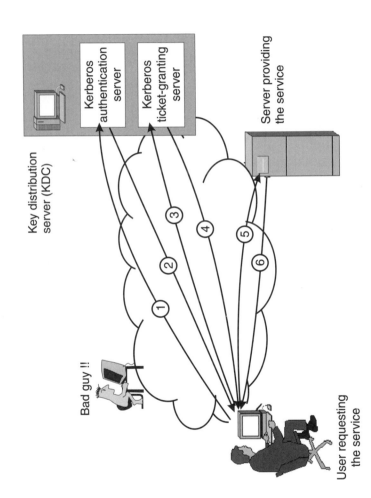

Figure 5.10 Kerberos authentication scheme.

2. The AS issues the required credentials based on the profile of the user in its database. The profile consists of the user ID and the associated level of authorization to access various services on the network. The credentials are sent encrypted over the network using a key based on the user's password. The AuthS server also issues a secret key that it shares with the user and the TGS server. The KDC stores passwords for all users on the network. This introduces a potential security risk if the KDC is ever compromised.

3. The user decrypts the credentials using his or her password. Please note that during this conversation no password was transmitted over the network, thus the potential hacker is unable to get hold of the password. This is the beauty of Kerberos. The client on the user workstation forwards this information with a request for a "service-granting ticket" to the TGS. The request is encrypted using the secret key received from the AS server.

4. The TGS decrypts the information using the secret key it received from the AuthS server. It then verifies the user credentials and issues the requested service-granting ticket for any server within its "realm," or domain of control. The ticket is encrypted with the service password.

5. The user, now armed with the required ticket, sends a service request to the service-providing server.

6. The service on the server decrypts the information using its own password and grants the services to the user.

The user has to go through this process for every new service. Each service-granting ticket is valid for eight hours. During this time the user does not have to repeat this process for the authorized service.

Pros:

- No password information is sent over the network in clear.
- The user is given access to services based on his or her profile. Each "Kerberized" application can control the user access based on the profile.

Cons:

- The major drawback of the Kerberos protocol is that you have to Kerberize every service on the network. This means modifications on the client side and the server side of each application. This could be a major undertaking and needs the source code of the application.

- The Kerberos server, hosting the AS and TGS, is one point of failure. If this server dies or becomes unavailable on the network all the network services become unavailable.

- The Kerberos server is also the most critical server since it holds a database of all user passwords. If this server gets compromised the whole Kerberos environment is compromised.

- The user workstation has to be a single-user workstation. If the workstation is a multiuser workstation, another user on the server can steal the ticket.

- There may be cases where the user workstation itself has been compromised before the user even logged into the workstation.

- You can find more information on the Kerberos at http://web.mit.edu/kerberos/www.

Digital Signatures

We saw the operation of public-key cryptography in the section on data confidentiality. Let's see how public-key cryptography provides authentication. Please refer to Figure 5.11 for the rest of the discussion.

When User A wants to send an authenticated and encrypted message to User B using RSA cryptography, the following things happen:

- *Step 1:* Each user, A and B, creates a private-key/public-key pair.
- *Step 2:* When a message is to be sent from User A, User A first encrypts the message using his or her private key A_{pri}.
- *Step 3:* The whole message is then once again encrypted by User B's public key, B_{pub}, and sent across the network.
- *Step 4:* When the message is received by User B, it is first decrypted using his or her own private key, B_{pri}.
- *Step 5:* This message is then decrypted again by User A's public key A_{pub}.

How does this operation provide authentication? When User B has decrypted the message in step 4, he or she is presented with another encrypted message. If the message was indeed sent by User A, the message must have been encrypted by User A's private key. Since there is only one public key corresponding to the private key of User A (each private/public key is a unique key pair), User

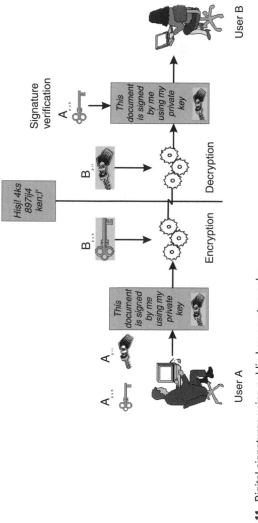

Figure 5.11 Digital signatures using public key cryptography.

B can use User A's public key to decrypt the message (step 5). If this step is successful, we can confidently say that the document indeed came from User A. In step 2, when User A encrypts the message, he or she is considered to be "signing" the message using his private key, thus providing a "digital signature."

Note   If the message body is a large message, the process of encrypting the document twice is a time-consuming and a resource-intensive process, resulting in a performance penalty. To avoid this performance penalty, when User A signs the message, he or she usually signs a hash (MD5 or SHA) value of the message and not the entire message. As we have seen in our previous discussions, the hash value of a message of any size is a small 64-bit or a 128-bit value. The performance penalty for encrypting the small hash value is minimal. This operation has the added benefit of providing a data integrity check to the message exchange.

Digital Certificates

In the previous section we talked about digital signatures; we will continue this discussion with an introduction to digital certificates, certifying authority (CA), and public key infrastructure (PKI). We will also briefly discuss the directory requirements. The next section on single sign-on will unify all these concepts for centralized security management within an intranet or an extranet.

What Are Digital Certificates?

Digital certificates are digital documents that provide a definitive association between a public key and the owner of the public key. The owner of the public key could be an individual, an application, or a service. Digital certificates try to solve the following problem.

In our previous example (Figure 5.11), User B uses User A's public key to verify User A's signature on the document. How does User B know that the A_{pub} is indeed User A's public key? If the document is a forgery, User A's private key signature, A_{pri}, and the associated public key, A_{pub}, could be forgeries as well! There are a couple of ways to verify the User A keys. If User B knows User A's voice, User B could call User A over the phone and verify the fingerprint of A_{pub}. Assuming that it is indeed User A on the other line (and not an impersonator), the key has been verified. This method of key confirmation works in an environment with small groups of individuals. What if we are talking about communicating with thousands of unknown individuals over the Internet or if the communicating entities are not individuals but computer programs? There needs to be a better way to provide the verification of a public key. This method is digital certificates.

Who Issues the Digital Certificates?

Digital certificates are issued by a certifying authority. The CA issues and vouches for the digital certificates by signing the certificate with its own private key. Anyone can verify the authenticity of the CA signature by decoding it with the CA's public key, much like the example in Figure 5.11. Why should you trust the CA? There is no clear answer to this question. At some point you have to trust somebody! Most CAs back their signatures with liability protection, but this amount today is pretty small. For a higher level of confidence, the CA issuing the certificate could be certified by another CA in the same hierarchy, thus establishing a "chain of trust." The whole e-commerce model on the Internet is dependent upon this chain of trust (see Figure 5.12).

The regional or country CAs could be maintained by private corporations or government organizations. Today, the CAs are mostly private corporations, like Verisign and Entrust, although United States government organizations like the Postal Office are planning to get involved in certification business. In the case of an intranet or an extranet, the regional CAs could be the regional

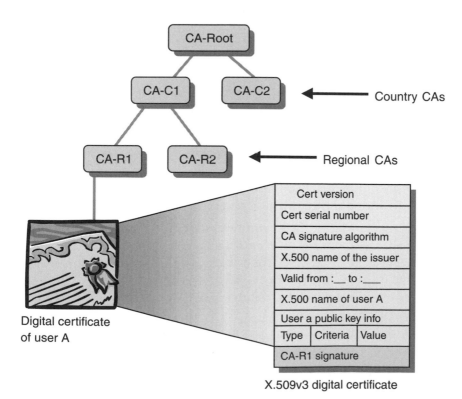

Figure 5.12 CA chain-of-trust.

security groups within corporations who are certified by the central corporate security group. For a multinational, there would be multiple country-specific CAs who are then certified by the "Root-CA" at the headquarters, thus creating the "chain-of-trust" within the organization CAs.

It is much easier to identify and certify a legitimate employee within an organization than an individual on the Internet. Corporations could choose to give certificates to their employees based on their role within the organization and areas of responsibility. In the case of X.509 version 3-based certificates, the policy-based information is reflected in the (type, criteria, value) field within the certificate (see Figure 5.12).

In the case of individual certificates, the certificates are suppose to be tied to the individual. It is the individual's public key that is being certified by the CA. What happens when the user leaves his or her desktop or a laptop for an extended period of time and there is no screen lock? Since the individual's private key must be on the desktop or the laptop, any other individual can potentially impersonate the valid user and conduct malicious activity using a valid user's key. How can one protect against such impersonation? The only method to protect against such intentional or nonintentional events is to have the private key located on a portable smartcard. If the desktop or the laptop is fitted with a smartcard reader, the smartcard can be inserted or removed as one wishes, protecting the private key. New portables from IBM are expected to be equipped with smartcard readers. Of course, if the user leaves his or her smartcard in the reader when he or she is away from the desktop, nobody can protect the private key. The only way to prevent this event from happening is to constantly educate the users on the security implications.

There are different types of digital certificates.

- *Client certificates:* These certificates are issued to individuals. These certificates are also known as personal certificates.
- *Server certificates:* These certificates are issued to application servers such as Web servers. The certificates are tied to the fully qualified domain name of the server. For example, www.amazon.com. If you change the name of the server in any way, you will have to issue a new certificate. An example of the Amazon secure Web server certificate is shown in Figure 5.13.
- *Software publisher certificates:* These certificates are used to authenticate software code or a software download from an ftp server.

The server certificate in Figure 5.13 shows that it has been issued by CA RSA Data Security, Inc. to Amazon.com, Inc. for the server address

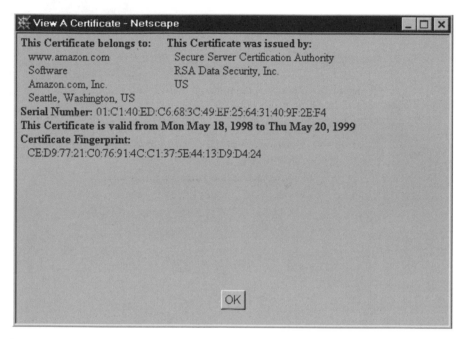

Figure 5.13 Server certificate issued by a CA.

www.amazon.com. The certificate has a unique serial number with the time-frame for which this certificate is valid. A certificate fingerprint is given at the bottom for verification. You can observe certificates of a secure Web server by connecting to it using Netscape or a Microsoft browser. In Netscape Communicator 4.0x, you can get to the security menu via Communicator → Security Info → View Certificate. In Internet Explorer 4.0, you can get to the same information by following File → Properties → Certificates, when you are viewing a secure document.

Digital certificates are currently used by secure socket layer (SSL) protocol, S/MIME, single sign-on solutions, PKI, and many other Internet applications. In an SSL connection, any browser capable of supporting an SSL protocol connects to a secure Web server (https) over a TCP/IP network. The client and the server could be on the Internet, within an intranet, or within an extranet security domain. The communications between the two entities is shown in Figure 5.14.

This message exchange is possible because the digital certificate of the CA signing the Web server's digital certificate is embedded in the browser. When Netscape first started with Netscape Navigator, it bundled RSA Data Security's CA certificate in the browser. Over the evolution of the Web and the browser, the latest version of the Netscape browser includes almost 35 CA certificates.

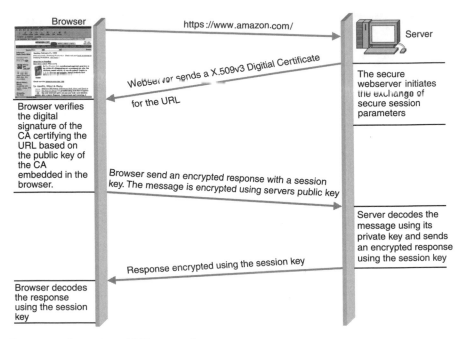

Figure 5.14 Operation of SSL protocol.

You can see the list of embedded CA certificates via Communicator → Security Info → Signers (see Figure 5.15).

When a browser receives the server certificate, the first thing it checks is the validity of dates in the certificate. If the dates have expired or are not yet applicable, it gives out an error message. Next, it checks the name of the server that has been certified by the CA. If this matches the URL it is trying to connect, the browser sends an OK response as shown in Figure 5.14. If the certifier CA key is not embedded in the browser, the browser can download the CA key. Since the new CA is not reliably known (embedded) to the browser, it will generate a caution message to the user about the identity of the Web server.

Public Key Infrastructure

Once you have thousands of users of public key cryptography and the digital certificates, it becomes essential to have a key-management infrastructure that can perform following functions:

- Manage the issue and revocation of digital certificates;
- Provide key management including key backup, key update, key recovery, and key escrow;

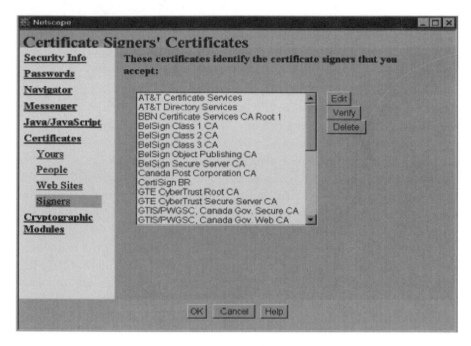

Figure 5.15 Embedded CA keys in Netscape browser.

- Provide security policy management including key life cycle and key usage.

Components of PKI infrastructure are shown in Figure 5.16.

We have already discussed the digital certificates and the CA component. Let's take a look at the directory component and the single sign-on component.

Directory

The most demanding component of a PKI infrastructure is the key-management component. When you have thousands of users with multiple keys and you need to manage the total key life cycle, the job becomes very complicated. Add to this the tasks of key backup, key update, and recovery of lost keys and you will have your hands full. As if this is not sufficiently complicated, think about a national or a global enterprise with intranet and extranet users. The task seems almost impossible!

But do not despair, there may be a solution. The solution is in the form of a centralized global directory. The directory could be based on a standard X.500 directory, LDAP directory, or other metadirectories. There are vendor products now in the market that will allow you to consolidate various application directories from Windows NT, UNIX, MS Exchange, Lotus Notes, cc:Mail, databases, and other applications into one central directory. The directories could be replicated or set in a master/slave arrangement with all entries synchronized. This allows administrators to get their hands on all the user-related information, including but not limited to their login IDs, passwords, storage space, and mail IDs. Security administrators can centrally create new accounts, modify old accounts, or delete accounts of employees or contractors who have left. The metadirectories can hold additional information like employee authorization levels, digital certificates, personal and departmental information. This simplifies the life of a security administrator since he or she does not have to run around fixing accounts, adding users in ten different systems, or trying to make sure that all access from any system on the intranet or extranet has been deleted for a departing employee or a contractor. Authentication can now be centralized at the directory.

There is a lot of activity in the directory market. Some of the major vendors of directory and meta-directory products are Zoomit, Wingra, Microsoft, Novell, ISOCOR, ICL, and Siemens. Zoomit has most recently been acquired by Microsoft.

If you add a centralized password and security administration to the central directory, you have just fulfilled the wishes of the administrator and re-

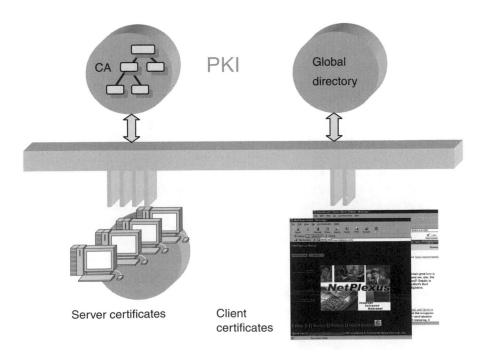

Figure 5.16 Public key infrastructure components.

duced security holes in your network. How about the user? Can we do any thing to simplify his or her life? Currently, if the user has to access 10 to 15 different accounts, he or she needs to have individual user IDs and passwords. He or she needs to remember them, maybe change them on a monthly basis, and follow strict guidelines for selecting a new password that is not too simple to crack and not too complicated to remember. Since a complicated password will lead to all the security vulnerabilities mentioned earlier in this chapter, the answer to a prayer is the emerging field of single sign-on.

Single Sign-On (SSO)

The single sign-on (SSO) solutions try to address the following needs within organizations:

- Enable users to access all the authorized applications with a single login name and password;
- Allow central administration of all the security policies, account management, and reporting;
- Encrypt the network communication within the intranet or over the extranet. The network communication includes application login and data communication.

When combined with one-time passwords, the SSO provides a formidable login mechanism for remote application and network access. Most SSO solutions in the market are based on a DCE (distributed computing environment) model and use Kerberos to provide security. The DCE model consists of a central authorization and security server with DCE agents on each desktop (see Figure 5.17).

In most SSO solutions there is very little change to the application, though you may have to write custom modules interfacing the client side of the application to the SSO API providing the seamless SSO.

How Does SSO Operate?

Consider a scenario of a corporate intranet with multiple NT domains. Some NT domains have trust relationships with each other, while others do not. When a user logs into his or her desktop for the first time on any particular day, he or she normally logs into the NT domain he has been assigned to. After he or she completes the NT domain login process, he or she might log into other applications like Microsoft Exchange, Lotus Notes, intranet Web server, or telnet to a UNIX server during the course of the day. With SSO implementation,

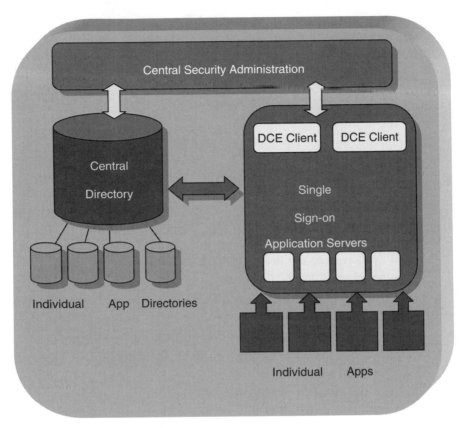

Figure 5.17 Architecture for global SSO.

the user has to complete only one login per day and is able to access all authorized applications without further logins. This is accomplished as follows:

- *Step 1:* When a user starts his or her desktop for the first time, the client SSO module on the user desktop intercepts the login and initiates a login to the SSO authorization server.
- *Step 2:* The user is unaware of this login to the SSO server and believes that he or she is logging into the NT domain server as usual.
- *Step 3:* The authorization server prompts the user for a password.
- *Step 4:* If the login is successful, the authorization server creates a cache of all the userID/passwords associated with authorized applications in the user's security profile. The cache is either stored on the authorization server or sent to the user's desktop.

- *Step 5:* At this point, the SSO module on the user's desktop initiates the NT domain login and supplies the userID/password from the cache.

- *Step 6:* The user is now logged into the NT domain.

- *Step 7:* Next, when the user double clicks on the "telent" icon on the desktop, the SSO module again intercepts the call and supplies the userID/password parameters to the application and logs the user into the remote UNIX system without him going through a login process.

A similar process is repeated when the user logs into other applications from the desktop.

The SSO login process is transparent to the user presence in the network (i.e., if the user logs into the network from a remote branch office, instead of his or her desktop, he or she still has to deal with only one login).

The following vendors are the leading vendors in the SSO market:

- Intellisoft with their Snareworks product line;
- Bullsoft with their OpenMaster Suite;
- IBM with the Tivoli framework and Global Sign-on product.

Other vendors like Axent, Hewlett Packard, and Computer Associates have a specific solution for a specific aspect of SSO and security management.

Note SSO offers significant ROI if you consider an enterprise with 10,000 users and make the following calculations:

Number of users (U) = 4,000
Average number of logins that a user performs in a day = 4
Average cost of user time per hour = $57.00

Assumption: We assume the average salary to be $80k per year with 50% overhead ($40k). The number of working days in a month are assumed to be 22 days with each day of 8 hours. That is: $80k + $40k = $120k per year per 12 months = $10k per month = $10k per (22 days × 8 hr per day) = $57.

Average amount of time spent by a user on logging in per year = 17.6 hours.

Assumption: Our assumption is that the user spends on an average 60 seconds logging into various applications. This translates into (4 logins) × (60 sec/day) × (22 days/month) × (12 months) × (1/3,600 sec/hr) = 17.6 hours.

Annual cost saving due to SSO = 4,000 users × $57/per hr × (4 logins − 1 SSO login)/4 logins × 17.6 hr = **$3,009,600!**

Approximate cost of SSO implementation = $600,000

Net savings within the first year of implementation = $2,409,600

If you are evaluating SSO product offerings, consider the following questions for your request for information (RFI) or request for proposal (RFP):

General Characteristics

- *Q.1:* Is the SSO implementation invasive in any way on the client or server side of the application?
- *Q.2:* What are the supported client platforms (e.g., Win3.1/95/98/NT, UNIX platforms)?
- *Q.3:* What are the SSO server side requirements (e.g., the OS versions, memory and disk space requirement)?
- *Q.4:* Does the SSO solution support one-time passwords including token (software/hardware) authentication?
- *Q.5:* What standards does the SSO support (e.g., DCE, SESAME, Kerberos, GSS-API)?
- *Q.6:* What is the storage mechanism for the information stored on the SSO server (e.g., RDBMS or proprietary)? Is the information encrypted when it is stored?
- *Q.7:* Is the userID/password information being cached? If yes, what is the "aging" mechanism for expiring the cache?

Nomadic Use

In the world of networked resources on the intranet or an extranet, it is important that users should be able to access an authorized information resource from anywhere in the network. In order to accomplish this task the following features of SSO products are essential:

- *Q.1:* Can a client desktop in a common area be shared by many end users? If yes, how can the two separate users and their associated userID/password information coexist without a security risk?
- *Q.2:* Given that Windows95 and Windows NT rely on locally stored passwords for authentication, is there a conflict between SSO login and the locally stored password?
- *Q.3:* Are the date and time of the last successful sign-on clearly shown in order to highlight potentially unauthorized sign-ons?
- *Q.4:* Is the name of the logged-in end user prominently displayed to avoid inadvertent use of client desktops by other end users?

Authentication

Authentication ensures that transactions are in fact initiated by the authorized end users. For security audit purposes, recording of all significant end-user activities (e.g., attempted login, activated applications) in a database is highly desirable. Check on the types of authentication methods supported:

- *Q.1:* Username and password?
- *Q.2:* Challenge handshake authentication protocol (CHAP)?
- *Q.3:* S/Key?
- *Q.4:* Digital certificates?
- *Q.5:* Token cards?
- *Q.6:* Challenge-response authentication method (CRAM)?
- *Q.7:* User defined?
- *Q.8:* Does the authentication server log all sign-on attempts?
- *Q.9:* What are the reporting capabilities of the SSO solution?
- *Q.10:* After a site-specified number of sign-on attempts, can all sign-on attempts be unconditionally rejected (and reset by administrator only)?
- *Q.11:* Is an inactivity timer available to lock and/or close the desktop when there is a lack of activity for a period of time?
- *Q.12:* Can the desktop be easily locked and/or closed when someone leaves a client desktop?

Encryption

Encryption feature ensures confidentiality and integrity of the network traffic between the client (user/application) and the security server(s), as well as the client (user/application) to the local/remote applications.

- *Q.1:* Is all traffic between the client and the SSO server encrypted?
- *Q.2:* Is all traffic between the client and the end applications encrypted? Does it need any change to the client or the server side of the software?
- *Q.3:* What are the cryptographic algorithms supported by SSO (e.g., MD5, SHA, DES, 3DES, IDEA)?
- *Q.4:* What are the authentication and key exchange methods supported by SSO (e.g., Kerberos, X.509, Diffie-Hellma, DSS)?

- *Q.5:* What is the overhead of SSO and associated security measures on the client applications/desktop?
- *Q.6:* What is the overhead of SSO and associated security measures on the SSO authentication server?
- *Q.7:* What is the overhead of SSO and associated security measures on the network?

Access Control

End users should only be able to access the applications that are configured in their profile. Indicate the access control parameters supported:

- *Q.1:* Source (IP address or host name) and port;
- *Q.2:* Destination (IP address or host name) and port;
- *Q.3:* User identity and/or group affiliation;
- *Q.4:* Time, day, and/or date;
- *Q.5:* Application and/or service;
- *Q.6:* Authentication method and/or encryption algorithm.

Application Level SSO Control

- *Q.1:* What mechanism is used for secure access to the mainframe applications (e.g., proxy gateway)?
- *Q.2:* What types of application or protocol adapters are available to automate the application launch process without having to adjust the individual applications? Are the adapters API-based, OLE-based, DDE-based, or scripting-based?
- *Q.3:* Are all application activations and de-activations logged?
- *Q.4:* When application passwords expire does the SSO product automatically generate new passwords? If yes, how are the administrators and users notified? Is there a new password registration process within SSO?
- *Q.5:* Are inactivity timers available to terminate an application access when there is a lack of activity for a period of time?
- *Q.6:* Does the SSO product log end users' activities within an application that requires multiple levels of login?
- *Q.7:* What are the development tools provided with the SSO product suite?

- *Q.8:* Are the SSO APIs open, published, and stable?

- *Q.9:* What is the level of expertise needed to integrate custom applications with SSO using APIs and SDKs? This criterion is very important since this will the biggest hidden cost of SSO implementation.

- *Q.10:* How does SSO interoperate with firewalls? Any known issues with major firewall vendors like Checkpoint, Cisco, and Gauntlet?

Security Administration

The power of the administration tools is key since the cost of administering a large population of end users can easily overshadow the cost of the SSO product itself.

- *Q.1:* Does the SSO product allow for the central administration of all userID/passwords of the end systems and applications?

- *Q.2:* How does a central change to a user profile get propagated to the remote applications/servers? Is the process automated or manual?

- *Q.3:* Is the user account administered at the end user level or at a group/role level?

- *Q.4:* Is it possible to install the SSO client remotely? Please indicate your confidence level in this operation (very high, high, OK).

- *Q.5:* Is the SSO client downloadable and dynamically configurable using client browser and Web server software?

- *Q.6:* Indicate the type of templates available for account administration.

- *Q.7:* Can you bulk-load and activate/deactivate large number of users?

- *Q.8:* Is dynamic user registration/modification/deletion supported on the SSO server?

- *Q.9:* Does this operation require any server reboots?

- *Q.10:* What GUI tools are available to change the end users profiles?

- *Q. 11:* Does the SSO server administration utility provide different levels of administrative access and the associated screens? This allows a certain level of delegation of responsibility among security administrators in various regions or departments.

- *Q.12:* Can an SSO server administrator list the number of active end users and the requested application?

- *Q.13:* Can events on the server trigger SNMP traps, output to a syslog, send an e-mail, or be piped to a custom application?

- *Q.14:* Do you support dynamic Web-based report generations capability?
- *Q.15:* Do you support command line administration of SSO server? Are there any standard command line utilities?

Reliability and Performance of the SSO Product

Given that an SSO product is positioned between the end users and the applications they need to access, it has high visibility and hence needs to provide a reliable solution with high availability and load-sharing capabilities.

- *Q.1:* Does the product provide fault-tolerance (cluster) and global-replication mechanisms? What is the cost of such a solution?
- *Q.2:* Does the product provide policy-based dynamic load balancing under master/slave configuration?
- *Q.3:* Check on the maximum number of simultaneous client connections supported by the SSO server.
- *Q.4:* Check on the Year 2000 compliance status of the SSO product suite. This should be a mandatory requirement. Obtain appropriate documentation from the vendor.

Cost of the SSO Solution (Including the Cost of Training and Support)

- *Q.1:* Check on the cost structure for *X* number of user licensees.
  - Software licenses cost;
  - Annual support cost;
  - Cost of hardware required.
- *Q.2:* Is the product extensible for custom applications?
- *Q.3:* What educational services are offered for the product? And the cost of training:
  - End users;
  - Programming;
  - Administrations.
- *Q.4:* Check on the level of post-sales help desk and technical support available for the vendor product (e.g., 7 × 24 technical support, on-line self-help site, bug fixes, updates).

Vendor Information

- *Q.1:* Company revenue for last fiscal year?

- *Q.2:* Number of years in business?
- *Q.3:* Number of employees to support the SSO product?
- *Q.4:* Availability of local technical staff in your area?
- *Q.5:* Other geographical locations (sales and support)?
- *Q.6:* Industry concentration of the vendor?
- *Q.7:* Participation in standards-development organizations?
- *Q.8:* Check on the total number of active customers and average number of users per customers in the install base of the product.

PAP/CHAP Authentication for Remote Access

Providing remote access to the information resources on the intranet or an extranet is a common business need today. We have looked at the inner workings of point-to-point protocol used for the asynchronous dial-up access in Chapter 2. In this section we will consider in detail the authentication phase in PPP-link negotiation (see Figure 5.18).

There are two types of PPP authentication protocols:

1. PAP (password authentication protocol);
2. CHAP (challenge handshake authentication protocol).

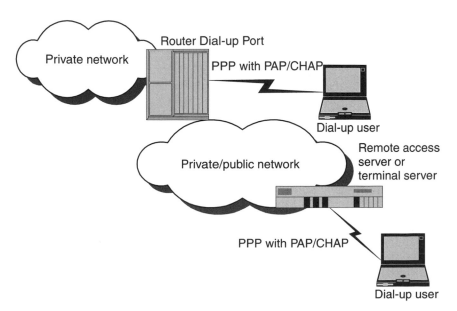

Figure 5.18 Dial-up authentication for PPP.

PAP

PAP is the simplest of the two PPP-authentication protocols and the least secure of the two as well. PAP authentication uses a simple user ID/password combination that has been preprogrammed into the dial-in client. The authentication information is sent in clear and can be easily tapped. PAP authentication is highly favored over CHAP by most Internet service providers today due to its simplicity.

PAP is a simpler authentication scheme than CHAP. In the case of PAP, the dial-in client is configured with a PAP user ID and password. The dialed router or the terminal server is also configured with the same user ID and password. The dial-in client sends its user ID and password continuously to the dial-in server until the call is accepted or terminated. The typical configuration commands on the router are shown in Figure 5.19. The PAP authentication information travels in clear over the network, hence is prone to snooping. One way to prevent a snooping attack is to use one-time passwords such as the token passwords. The RADIUS servers authenticating the dial-in users can be programmed to authenticate against a token authentication server like the ACE server from Security Dynamics.

CHAP

CHAP provides a stronger authentication scheme for dial-in access. In the case of CHAP, the dial-in client and the dial-up server share a common secret key, (K_{pri}), just as in symmetric key encryption. This key is given to the client in the setup diskette or by separate means. Please refer to Figure 5.20 for operation of the CHAP protocol.

- *Step 1:* When the PPP-link negotiation process enters the authentication phase, the dial-in server issues a challenge to the dial-up client. The challenge consists of the server name, an ID (like a sequence number), and a random number, *r*.

Router # conf terminal

Router (config) # username john password trustme

Router (config) # interfac async 1

Router (config-int) # encapsulation ppp

Router (config-int) # ppp authentication pap

Figure 5.19 Cisco router configuration for PAP authentication.

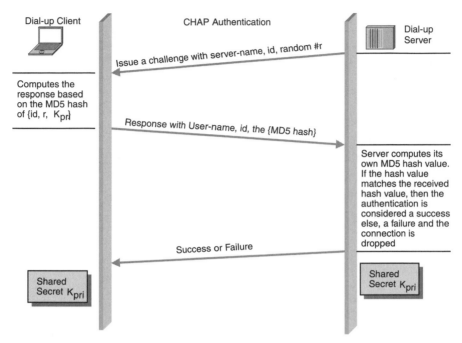

Figure 5.20 Operation of CHAP authentication scheme.

- *Step 2:* The client, upon receipt of the challenge, computes a response consisting of the userID, the ID sent by the server, and a MD5 of the ID, random number r and the private key, K_{pri}.
- *Step 3:* The server computes its own hash value using the same parameters. If this hash value matches with the received hash value, and the ID matches the ID in the challenge, the server sends an "authentication success" response to the client (provided the user has an account on the server). In case of a failure, the connection is immediately dropped and an error message is sent to the client.
- *Step 4:* Upon successful completion of the authentication phase, the PPP link negotiations enter the NCP phase.

The biggest security risk for the CHAP protocol is that the secret key, K_{pri}, is stored on the client laptop as a hidden file. Unless you physically secure the laptop at all times, there is a danger that the key could be easily stolen. There is also a danger of key exposure during the initial key exchange.

The typical configuration commands on the router acting as a dial-up server are shown in Figure 5.21.

Router # conf terminal

Router (config) # username john password trustme

Router (config) # interfac async 1

Router (config-int) # encapsulation ppp

Router (config-int) # ppp authentication chap

Figure 5.21 Cisco router configuration for CHAP authentication.

In CHAP, even if somebody captures the session packets the information is not very useful since the key is unique for every session.

5.5 Firewalls

Firewalls play an important role in the security of both intranet and extranet network infrastructure. In this section we will explore different technologies for implementing firewalls, various vendor products, and their pros and cons.

5.5.1   Origin of Firewalls

In the medieval times the concept of a firewall was evident in the architecture of castles and the townships around castles (see Figure 5.22).

The main part of the castle was surrounded by a tall wall and maybe one or two gates. The gates were always guarded and the flow of goods and visitors was carefully screened by the guards at the gates. In order to protect the township around the castle another wall was built around the township with its own gates. Although the gates had guards, the traffic control at these gates was not as strict as at the gates on the inner wall. The walls and gates provided certain levels of protection against invaders or other malicious elements in the surrounding area.

This theory of protection by isolation and controlled traffic flow has sustained over a long period and is evident in the network and security architecture of a modern day intranet and an extranet (see Figure 5.23).

5.5.2   What Is the Role of a Firewall in an Intranet or an Extranet?

In today's rapidly expanding networked enterprise, a firewall provides the necessary isolation between an external and an internal network. In the case of an intranet, the network isolation in today's intranet is provided by controlling the

Figure 5.22 Perimeter defense architecture.

network connection(s) to external network(s) through firewalls and routers, while the guarded traffic-flow functionality is provided by the policies and access control rules on the same two devices. Different layers of firewalls provide various lines of defense in order to protect the valuable resources and information assets on an intranet and/or an extranet. The external network in this case could be the Internet or another private network, while the internal network is the corporate intranet consisting of many local and distributed LANs connected over a WAN. With the increasing threat from hackers, crackers, and joy riders, firewalls are an essential part of any intranet or extranet architecture. A firewall, as a physical device, could be a router, a dedicated black box, or a server based with a software firewall (e.g., Checkpoint Firewall-1 and Gauntlet).

In addition to controlling the traffic, firewalls also provide additional capabilities such as:

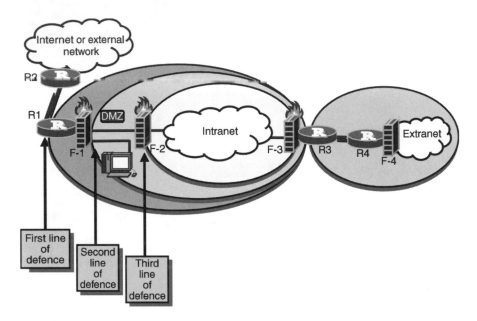

Figure 5.23 Perimeter defense using firewalls.

- Logging and ability to conduct audits;
- Ability to set up VPN connections over a public or a private network;
- Traffic director and load balancing. Due to the fact that the firewalls are located at strategic points in the network, new hardware-based firewalls provide additional capabilities to direct traffic over the least-used path.

5.5.3   What Are the Different Types of Firewalls?

Firewalls can be categorized based on the technology used to protect the internal network:

- Packet-filtering firewall;
- Proxy-based firewall or application firewall;
- Firewall based on stateful-inspection.

Packet-Filtering Firewall

A packet-filtering firewall is usually a router-based firewall. Let's consider the routing process a router goes through while routing an IP packet. A similar concept applies to IPX or Appletalk packets.

As a router receives the packet, it examines the header of the IP packet to gather information like:

- The destination IP address;
- The source IP address;
- The destination TCP or UDP port number;
- Type of protocol (IP, TCP, UDP, or ICMP);
- The message type (SYN, ACK, FIN, RST).

Based on this information and the internal routing table of the router, the router determines the destination port where the packet needs to be sent so as to route it to its ultimate destination.

Based on these two steps and the position of the router within a network (e.g., router R1, R3, and R4 in Figure 5.23), the router is in an excellent position to apply forwarding rules to every IP packet that is traversing the router. For example, the rules could state that:

1. All packets to a destination address be blocked;
2. All packets to a destination address be blocked except for a certain range of IP addresses;
3. Allow all packets going to specific TCP ports like Web (port 80), DNS (port 53), and SMTP (port 25) be allowed, while the rest of the packets be dropped;
4. Do not allow any packets from the external network with a source IP address of the internal network (this specific rule provides antispoofing protection).

This type of a device (router) providing the firewall functionality is known as a "packet-filtering firewall." In the case of Cisco routers, the rules are implemented as access control lists (ACLs). Each ACL has a following format:

Access-list {Access-list-Number} (deny\permit) {Protocol} {Source-ip-address of the host or thenetwork}
    {Destination-ip-address of the host or thenetwork} eq {application port}

Here is an example of a Cisco ACL:

access-list 131 permit tcp 207.80.128.0 0.0.0.255 any eq www

Depending on the applied interface and the traffic direction, this rule can have a unique effect or no-effect. If this rule was applied to the inbound serial0 interface of the router R1 in Figure 5.23, this rule can be interpreted as follows.

Permit a TCP request for Web connection from any host in the network 207.80.128.0 to any host in the internal network. The standard TCP port for Web connection is port 80.

Pros:

- It is quite inexpensive to build routers with packet-filtering abilities.
- The routers are already in the network providing routing functionality, thus avoiding additional cost of deploying a firewall.
- It is easy to configure and apply rules to the incoming or outgoing traffic on a particular interface of the firewall.

Cons:

- A router managing a high volume traffic site could get quickly overwhelmed when asked to perform additional packet-filtering functions, thus affecting the router performance and hence the network performance.
- The router connecting an external network to an internal network (router R1 or R3 in Figure 5.23) is potentially prone to a denial of service (DoS) attack, quickly isolating the internal network from an external network.
- A packet-filtering firewall is prone to mistakes in the configuration of the packet-filtering rules including the wrong TCP port or wrong sequence of rules, exposing the network for some types of known vulnerability.
- Packet filtering does not work well in an environment that needs dynamic rules. For example, routers cannot deal with TCP/IP services such as active FTP that use dynamic TCP ports at the source, or in the case of a DoS, the router filters need to be adjusted frequently to isolate a specific range of source IP addresses used by the attacker.

Note New generations of firewalls and routers have started to implement dynamic rules that detect a threat (like DoS) or a requirement (like active FTP) and create new rules on the fly.

- Routers cannot generate extensive audit logs.

- Packet filtering is a stateless process. Packet filtering cannot maintain any state of information of the preceding connection. For example, in the case of an active FTP connection from inside to outside, the returning data connection will be blocked unless random ports above 1023 are opened. This could lead to added security exposures on the higher number ports.

- Packet filtering does not provide an additional level of granularity based on the data content of the packet. For example, if the policy allows Web connections to specific URLs within a Web server, a packet filter will either allow or deny all traffic on port 80 to the IP addresses of the Web server. Our next firewall type overcomes this limitation.

Proxy-Based Firewall or Application Gateway

Unlike the packet-filtering firewall that operates at the layers 3 and 4, a proxy-based firewall operates at layer 7, the application layer, hence the name *application gateway*. The concept is very simple and can be explained from Figure 5.24.

The application gateway is a dual-homed (internal and external) host server (UNIX or NT) that is running proxy servers for each application needed by the end user on the internal network. The proxy servers handle the connection request from the internal clients to the external servers and vice versa. Each internal client is configured to point to the application gateway for access to a specific application like the Web, SMTP, or FTP. In Netscape, you can locate the browser configuration screen under Edit → Preferences → Advanced Proxies → Manual Proxy → View (see Figure 5.25). The operation of application gateway can be explained by following example.

Figure 5.24 Operation of application gateway.

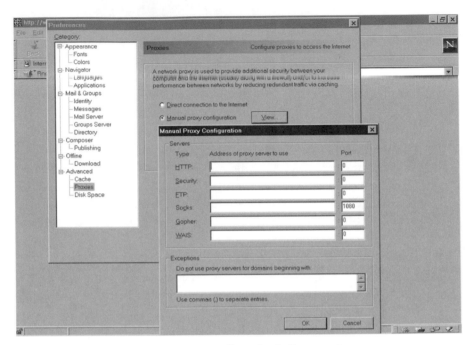

Figure 5.25 Application proxy gateway configuration in Netscape browser.

- *Step 1:* A user on the internal network requests access to a Web server on the Internet (external network).

- *Step 2:* Since the user browser has been configured to point to the application gateway as the proxy server, the request is forwarded to the HTTP proxy server on the application gateway.

- *Step 3:* Upon receipt of the HTTP request, the HTTP proxy server opens a new HTTP request to the destination Web server on behalf of the internal user (thus proxying for the user). This step is completed only if the request matches the requirements of the security or organizational policy configured on the server.

- *Step 4:* When the HTTP proxy server receives the reply, a new response connection is set up from the proxy server to the requesting user browser.

The key to the successful operation of a proxy firewall is the end-user transparency to the application gateways. Once the name and port of the proxy server have been configured on the client browser, the above steps occur transparently from the viewpoint of the client and the destination server.

Examples of proxy firewalls are:

- *SOCKS-based firewall:* The SOCKS firewall uses a generic proxy server and special SOCKS clients. All requests for connection to the external network are encapsulated using the SOCKS-defined encapsulation and are carried from the SOCKS client to the SOCKS server. The main disadvantage is that you need a special SOCKS-enabled client.

- *Firewall based on the TIS Toolkit:* Varies from the SOCKS concept of proxy firewall. The TIS Toolkit implements small proxy server for each Internet application, as explained in our discussion in this section.

- *Netscape and Microsoft proxy firewalls* are based on the proxy firewall principles discussed in this section.

Pros:

- Since the proxys are implemented at the application layer, unlike the packet filters, the state information of each connection is easily maintained. This allows an easy firewall setup for challenging applications like ftp.

- Since proxys are used to channel all the user traffic to the external network, they are a great place to apply organizational policies.

- The strategic location of a proxy firewall allows for implementation of network caching, thus potentially improving the performance of the connection. For example, if multiple users are requesting the same document, only the first user experiences any delay in getting the document. The subsequent users requesting the same document are provided with the document out of the proxy server's local cache. The modern caching servers have a tremendous amount of intelligence and ensure the "freshness" of the document.

Cons:

- It is hard to configure and maintain an unbreakable proxy server. Any changes to the OS configuration, proxy configuration, strength of the proxy code, or user accounts on the proxy server can weaken the firewall.

- As new applications are added to the user's needs, new proxy servers need to be developed for the application gateway. This could be a time-consuming and expensive process.

- Without the caching component and a proper management of the cache, there is a significant impact to the user perceived performance.

- Since all requests are channeled through the proxy server, the proxy server can be a single point of failure.

It is best to take a combination approach by combining the application-level proxy with a packet-filtering router.

Firewall Based on "Stateful-Inspection"

A third type of firewall employs a combination of application- and network-layer firewall functionalities and adds other modules like the encryption module. In the case of stateful-inspection, the inspection module in the firewall sits in the TCP/IP stack (the inspection module is located between the data link layer and the network layer of the OSI stack) of the firewall server. The module then decodes the IP packet header and the data portion of the packet for information on the source, destination servers, requested port, and portion of the content. The firewall then makes a table of all this information and adds a context-based information to the table. This context is used to evaluate each packet exchange and allows the firewall to manage not only TCP exchanges, but also the stateless UDP exchanges.

Checkpoint Firewall-1 is an example of stateful-inspection firewall. Cisco PIX firewall also incorporates context-sensitive firewalling features.

An example of Checkpoint Firewall-1 policy administration GUI is shown in Figure 5.26.

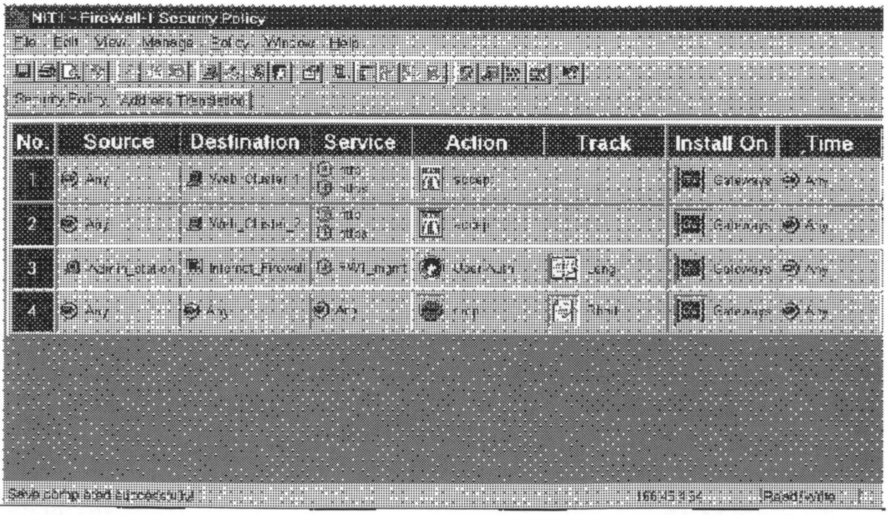

Figure 5.26 Security policy administration GUI for Firewall-1.

Pros:

- You do not have to develop separate proxy servers for every new, old, and current application.
- Since the firewall inspection and policy enforcement is happening at the network layer of OSI model, the firewall is very fast and efficient.
- No changes to the client software are necessary.

Cons:

- Since this is a software-based solution like all the previously discussed solutions, it has a performance limit based on the server hardware platform.
- Unlike proxy-based firewalls, Checkpoint cannot exercise control over content manipulation.

New Trends in Firewalls

As more and more functionalities like encryption, encapsulation, and load balancing are being added to firewalls, the performance of the software-based firewalls is likely to suffer. Some relief can be found with cheaper and faster hardware with increasing memory configuration. These hardware improvements are still not sufficient to bring wirespeed performance to the firewall operation. Another aspect of firewalls that is equally important is the administration GUI and the administration toolsets available for managing the ever-critical firewall configuration. As firewalls become common, the firewalls need to become idiot proof. They must operate like a blackbox or a security appliance. New hardware-based firewall products from Ascend, Nokia, Timestep, and Lucent allow you set up an idiot-proof firewall with wirespeed performance. Nokia and Checkpoint have introduced a layer 3 switch that incorporates Checkpoint's Firewall-1 technology in the switch hardware itself. Improved firewall hardware designs with ASICs for encryption and encapsulation allow you to get as close to the wirespeed as possible. The hardware-based firewalls also make it easy to set up a high-resiliency load-sharing environment. Some of the hardware-based firewalls have a footprint that is similar to a medium-sized modem. These smaller units can be deployed easily at remote branch locations or at homes. If you have a failure in any one of these units just replace it with a backup unit.

Network Address Translation

Most of the firewalls and routers are now asked to perform a network address translation function, also known as NAT. NAT allows organizations to continue to use private, unregistered IP addresses within the network, but still be able to connect to the public Internet and other extranet LANs with registered IP addresses.

There are two ways to use NAT:

- Perform one-to-one mapping of private internal addresses to publicly routable registered IP addresses.

- Map internal addresses to the single NAT address, with each internal IP address mapping to a unique port of the registered external IP address. How does NAT keep track of who requested what and how it can get back to the originator? The device performing NAT maintains an internal dynamic table of the request and response per session. New breeds of firewalls can accomplish this feat for both TCP and UDP sessions. Some examples of NAT devices are Cisco's PIX firewall, Ascend's Pipeline routers, and Checkpoint's Firewall-1 software.

VPNs

With the rapid growth in the Internet and its universal access, organizations have started to implement the corporate connections over the public network. The first step in making this transition possible is development of virtual private networks (VPNs), based on strong encryption and authentication schemes. In the next chapter we will see how we can use the VPN technology to set up a private intranet or an extranet connection.

5.6 Conclusion

There are many aspects of intranet and extranet security including authentication, authorization, confidentiality, integrity, and access control. Each aspect, when taken in the context of the three mantras of security policy—prevention, detection, and correction—form the basis for an enterprise-wide security plan. A strong and "known" security policy with a security plan that makes use of the best of the encryption, authentication, and data integrity tools will go a long way in ensuring the security of the corporate intranet and extranet infrastructure.

6

Virtual Private Network

Virtual private networking is once again a hot topic of interest among IT professionals. Last time around, the term *VPN* was associated with voice networks. Today, VPN is popularly associated with the data communications industry in general and the Internet industry in particular. In the case of voice networks, the VPN denoted a private network of voice circuit-switched connections carved out of the public telephone network for use by corporations. One such example of a voice VPN product is MCI Worldcom's VNET product line. In today's Internet-focused world, the term *VPN* serves a similar role to the one it played in the voice world. In this chapter we will take a look at:

- The definition of VPN in the world of data communication;
- Various drivers for the phenomenal growth of data VPN in the Internet, intranet, and extranet;
- VPN technologies for implementing intranet and extranet solutions;
- Leading VPN products and their implementations.

We will discuss sample implementations in this chapter and Chapter 7 on case studies.

6.1 What Is VPN?

VPN is a private data network carved out of a public data network like the Internet. Various techniques and technologies have made it possible to ensure the privacy of the data while it is flowing over a shared infrastructure.

VPN can be broadly classified based on its application (see Figure 6.1):

- LAN-to-LAN VPN for connecting various parts of an intranet;
- LAN-to-WAN VPN for extending an intranet to external entities to form extranets;
- Remote-LAN-access virtual private dial network (VPDN) for accessing intranet and extranet applications over a dial-up network.

6.1.1   LAN-to-LAN VPN for Connecting Various Parts of an Intranet

When a company wants to extend its internal network to remote regional or branch offices, an Internet-based VPN solution can provide significant cost savings. The intranet traffic is either encrypted or tunneled through the Internet using VPN technology to maintain data integrity and privacy. The traditional alternative of connecting by private lines is too expensive and does not scale very well. Other cost-effective means of creating VPN are based on layer 2 technologies like frame relay and ATM. We will investigate each one of these VPN technologies later in this chapter.

6.1.2   LAN-to-WAN VPN for Extending an Intranet to External Entities to Form Extranets

The ubiquitous access to the Internet is also ideal for extending the intranet to vendors, customers, and partners via extranet VPNs. VPN access can also be grouped together based on the extranet policy to create various communities of interests or closed-user groups (CUGs). The same VPN technologies used in creating intranet VPNs are used in extranet VPNs, but in a different context. Additional security and access controls can be added by introducing traditional firewalls (see Figure 6.1).

6.1.3   Remote-LAN-Dial VPDN for Accessing Intranet and Extranet Applications

The concept of ubiquitous access on a public data network has been borrowed from the world of PSTN. Just like the Internet, one can access the PSTN network almost anywhere in the world. VPDN provides dial access to intranet and extranet applications via the PSTN network. For example, VPDN provides the remote LAN access for mobile executives and employees who are telecommuting to work. The data traffic is secured by way of encryption and authentication.

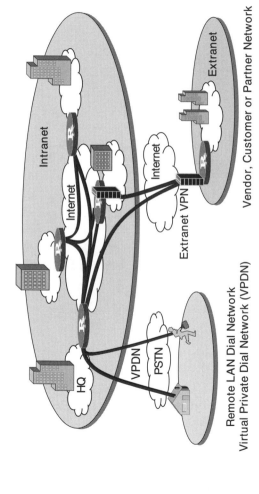

Figure 6.1 Types of VPNs.

6.2 Why VPN?

The preceding discussion gives us some clues to answer the question, "why VPN?" Let's explore this topic further. The various reasons for VPN can be summarized as follows:

- Lower cost of deployment;
- Data privacy;
- Ubiquitous access;
- Deployment flexibility;
- Implementation scalability.

6.2.1 Lower Cost of Deployment

The first and the foremost reason has to do with pure economics. If one were to build a network shown in Figure 6.1, using private lines, it would be quite an expensive proposition. The most cost-effective way to build an intranet or an extranet network is to leverage the carrier's investment in building the public data network like the X.25 network, the frame relay network, or the mother of all networks, the Internet. Due to the public nature of these networks, the corporate data traveling over these shared networks is not secured from hackers, crackers, and competitors. This leads us to the second reason for having VPNs.

6.2.2 Data Privacy

The VPN provides data privacy by either encrypting the data or tunneling the data over the public network. The tunneling mode is also useful in creating a multiprotocol internetwork. If we consider the public Internet, it is based upon the TCP/IP protocol. If you wanted to connect two LANs with SNA traffic, the SNA data will have to be encapsulated inside an IP packet and sent over the Internet using the tunneling mode of VPN. The VPN setup also adds authentication to both the end system and the end network.

6.2.3 Ubiquitous Access

The third reason is the "ubiquitous" connectivity of the Internet. It is now possible to get a LAN-to-Internet or dial-to-Internet connection in most parts of North America, Europe, and Asia (including Asia-Pacific). By accessing the network locally, organizations can save tremendous amounts of money. For example, a special case of VPDN setup allows an employee traveling in Europe to dial a "local" ISP number and be authenticated against the home network, thus providing tight security and cost savings due to the savings in long-distance charges. In Fortune 500 companies, these charges can quickly add up to multimillion dollar savings. Smaller companies can also take advantage of the ubiquitous Internet access to set up competitive operations, thus allowing them to compete globally with multinational competitors, all at the cost of a local connection. The small businesses can now divert the infrastructure dollars to acquiring new customers and providing the best service to the existing customers.

6.2.4 Deployment Flexibility

The fourth reason for the Internet-based VPN is that the VPN setup provides network-level flexibility. If the branch and regional offices are already connected to the Internet, they can now be added to the intranet at the click of a button. If the offices are not previously connected to the Internet, accessing the Internet connection allows these offices to take full advantage of connecting to the Internet as well as to the corporate intranet or the extranet. This dynamic nature of the connection setup within VPN provides any-to-any connectivity for corporate intranet or extranet setup. It is also useful in providing a high level of network and security resiliency.

6.2.5 Implementation Scalability

The Internet VPN solutions are highly scalable due to their simplicity and ease of deployment. The solutions are only limited by the processing power of the hardware. True any-to-any connectivity can be deployed without any scaling limitation.

In the next section we will take a look at various types of VPNs and their implementation technology.

6.3 VPN Implementation for Intranet and Extranet

The VPN implementation requirements are different for a network-to-network connection and for a dial-to-network connection. In either case, there are some fundamental selection criteria that must be considered to select the most appropriate technology. The criteria can be broadly classified as follows:

- Security;
- Performance;

- Ease of management;
- Conformance to standards and interoperability.

6.3.1 Security

The security aspect encompasses areas that allow us to maintain the data privacy, integrity and access control over a public network. The main areas are:

- *Data privacy:* The VPN solution needs to provide the highest level of data privacy, since potentially sensitive data is being routed across a public network. The two most popular solutions for data privacy are encryption, which means the data portion of the packet payload is encrypted using symmetric or asymmetric key encryption technology (the packet header is not modified), and encapsulation or tunneling. In tunneling, the entire data packet, including the header, is stuffed inside a new packet. This mode is used to transmit non-IP protocol over the IP backbone or IP-within-IP for security reasons (see Figure 6.2). In the case of encryption, the solutions need to support strong standards-based encryption technologies like DES, 3DES, RSA, IDEAS, while the tunneling needs to support standards like IPSec and L2TP.

- *Data integrity:* The VPN solution needs to maintain the integrity of data as it flows over the public network. Any changes to the mission-critical data by a wily hacker could prove to be disastrous for the company. The data integrity can be assured based on the secure hash algorithm (SHA), message digest (MD4) or MD5 hash algorithms. Cryptographic hashing uses a mathematical hash function to transform any variable-length input into a hash value of fixed size. The hash value is transmitted with the data and is used by the receiver to compare against its own hash calculations. If the hash values match the data, it is accepted, otherwise it is dropped.

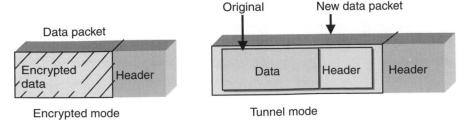

Figure 6.2 Data privacy by encryption and tunneling.

- *Authentication:* The VPN solution needs to provide support for various user-authentication schemes like the simple userID/password, token authentication, smartcards, or X.509v3 certificates.

- *Authorization:* The solution needs to provide a simple way to configure user profiles and their level of authorization. The changes to the authorization levels need to be logged and any violations need to be reported via some type of management notification or report.

- *Access control:* The VPN solution needs to provide as granular a level of access control as possible for both security and audit reasons. The access control could be based on userID, hostID, IP address, or a subnetwork address. The information on the user profile, including the time of access and user authorization credentials, needs to be stored in a central repository like an LDAP directory or an X.500 directory. A simple-to-use administrative GUI providing access to such a directory is desirable to manage the user-profile information.

- *Accounting:* The VPN solution needs to provide the means to gather accounting data on user activity for internal or external charge-backs.

6.3.2 Performance

One of the important selection criteria from the user perspective is the performance of a VPN solution. A user of a network-based service does not care if his or her access is over a private network or shared public network. The user would ideally like to have the same level of performance in either case. Many factors affect the performance of a VPN solution. For example, the Internet-based VPN has no predetermined path between any two end points. Hence, it is very difficult to estimate performance parameters like latency and user-response time. This is important in mission-critical applications based on SNA, which are extremely time sensitive, and could be affected by the unpredictable nature of traffic flow over the Internet. Additional overhead of encryption and encapsulation may add to the total processing time, adding to the latency and reducing the data throughput. If the VPN is confined to one Internet service provider network, it may be easier to get some assurances, even SLAs, on the performance characteristics of the VPN. If the VPN spans multiple service provider networks, similar performance SLAs are nearly impossible. In view of these challenges, it is important to consider the following performance parameters while designing or buying a VPN solution.

- *Quality of service:* The QOS is a complex parameter. It encompasses other parameters like end-to-end latency, dropped packets, and effective throughput. In the case of a VPN setup over the Internet, the only portion of the network under your control is at the edge of the network; the service provider controls the network backbone. In order to ensure a certain level of QOS, a VPN solution could be designed with bandwidth allocation and priority queuing mechanisms at the edge and in the backbone, to ensure priority for the VPN traffic. When these measures are combined with service level agreements from the Internet backbone provider, it is possible to achieve acceptable QOS goals. In the case of a frame relay or ATM-based VPN backbone, it is much easier to achieve the QOS goals due to the tight control over end-to-end design parameters.

- *Service level agreement:* Service providers have started to provide SLAs backed by some type of money-back guarantees for the end-to-end traffic, provided the traffic stays on their backbone. Some providers have gone to the extent of creating a parallel IP backbone for VPN services alone. Where there is no parallel network, most service providers have spent enormous amounts of money constantly upgrading their backbone to keep up with the traffic demand. Most tier 1 service providers maintain 40% to 50% of buffer capacity in their network. Almost all the tier 1 network operators like MCI Worldcom, Cable & Wireless, GTE, and Sprint maintain Web sites that show real-time measurement of the latency and packet loss in their backbone. Third-party performance monitors like Boardwatch provide an independent measurement of the backbone performance.

- *Multiprotocol support:* The VPN solution needs to accommodate legacy networks and protocols like Novell's IPX and IBM's SNA protocol. The most predominant method to transmit the legacy data over the IP backbone is to encapsulate it. The process of encapsulation and decapsulation adds to the processing time on edge devices like routers and firewalls, which can affect the total performance.

- *Reliability and resiliency:* Design considerations for building high-resiliency VPN networks have a significant impact on the ultimate performance, availability, and cost of the VPN service. It is prudent to identify all the single points-of-failure and build a high-resiliency architecture that can operate with one or more failures.

6.3.3   Ease of Management

- *Central security and policy manager:* The VPN solution needs to provide for a central security and policy manager in an easy to use graphical user interface. The cost of VPN administration can add up very quickly and equal the cost of the VPN solution itself. Another way to look at this important aspect is the indirect cost of a security breach if the VPN security has not been properly configured. A central policy manager should be able to set user-profile parameters like:

- The access time, Monday to Friday, 8 A.M. to 6 P.M.;

- User and group authorizations to specific servers and files;

- Authorized source and destination networks;

- Setting of userID and password credentials including add, modify, and delete operations;

- Enforce authentication mechanisms based on OTP and tokens.

- *Address management:* The VPN solution needs to provide address management features when the networks being connected use private or unregistered addresses. Address management is also an issue when the dial-in users are given dynamic IP addresses by a local Internet service provider.

- *Event logging, auditing, and reporting:* An intranet or extranet implementation using VPN technology over a public network is susceptible to network intrusion, hacking attacks, and destructive viruses. It is important to be able to monitor the activities on the VPN in real time as well as in retrospect. Significant events based on programmed thresholds need to be logged and reported in ways that can aid in quick problem resolution and detection of any security breaches in the implementation.

6.3.4 Conformance to Standards and Interoperability

The VPN solution needs to conform to open standards as much as possible. Open standards provide interoperability and a choice of vendors in VPN implementation. Various open standards have been specified in the areas of encryption (IPSec), data integrity (MD5), proxy service (SOCKSv5), authentication (CHAP), key exchange (IKE, SKIP, Diffie-Hellman), and digital signatures (X.509v3). Additional de facto standards exist that have been championed by major vendors in the areas of operating systems (Windows), data tunneling (L2F, PPTP), and authentication (RADIUS, NT Domain authentication, TACACS+).

6.4 Network-to-Network Connection

As shown in Figure 6.1, a network-to-network connection could be used for connecting various intranet sites over a public network or extending the intranet connection to a vendor, customer, or a partner network. In this section, we will look at the LAN-to-LAN and LAN-to-WAN implementation of a VPN solution.

LAN-to-LAN or LAN-to-WAN VPNs can be implemented at various layers of an OSI stack (see Figure 6.3). We will examine the various technology options at each layer and explore their pros and cons.

6.4.1   Data Link Layer

The link layer VPN technology can be classified as:

- An overlay technology;
- Tunneling technology.

When a network layer protocol like IP flows over a frame or an ATM infrastructure, a predefined path already exists from node A to node B. This is a classic overlay model. The frame relay and the ATM technologies are predominantly used for LAN-to-LAN connection. The tunneling protocols like PPTP,

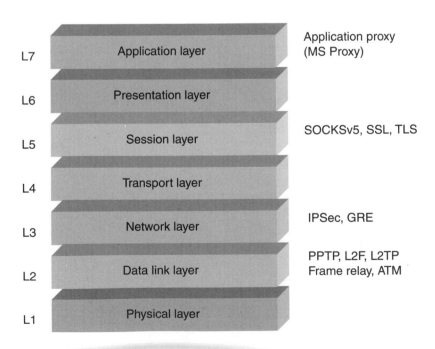

Figure 6.3 VPN technologies at various layers of an OSI stack.

L2F, and L2TP are hybrids of overlay and routing models, since the tunnel is mostly a point-to-point tunnel over a routed infrastructure. The PPTP, L2F, and L2TP technology is targeted towards dial-to-network connections. In this section we will concentrate on the frame relay and ATM-based implementation of VPN.

Frame Relay

We had an introduction to the frame relay-based VPN approach in Chapter 2. In this section we will summarize the approach, its pros and cons, and look at the implementation details by configuring frame PVCs on a Cisco router.

Frame relay VPNs are based on a shared-service infrastructure provided by existing telephone companies like MCI Worldcom, AT&T, Sprint, and British Telecom. The point-to-point connectivity is based on frame relay PVCs. Though this is a shared infrastructure, each PVC connection is isolated and the only common devices in the path of multiple end-to-end PVCs are the frame switches. Since frame relay operates at the data link layer, it can support any network layer protocols like IP, IPX, and SNA by using RFC 1490's "multi-protocol interconnect over frame relay" encapsulation method.

Figure 6.4 shows an intranet and an extranet frame relay implementation connecting the various regional and branch offices of an enterprise to a vendor network. The figure also shows the DLCI assignments for each PVC end point.

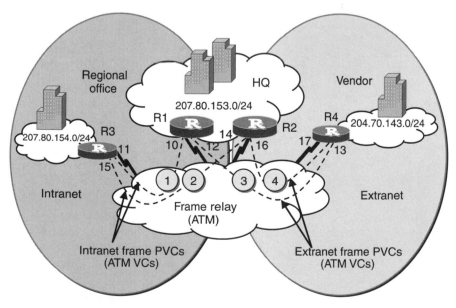

Figure 6.4 VPN setup using frame relay.

The intranet implementation with routers R1, R2, and R3, and two PVCs, PVC1 and PVC 2, provides a high-resiliency infrastructure for connecting the regional office to the HQ. The regional office could either load share the traffic across the two PVCs or use them in a redundant mode. An extranet implementation has been configured with frame PVCs three and four connecting the vendor network (R4) to HQ (R1, R2). Although both routers R1 and R2 share intranet and extranet PVCs, the traffic on each PVC is isolated. Each PVC terminates on its own subinterface on the router. Router access control lists can be added on each subinterface to further isolate the traffic on each subinterface.

What are the pros and cons of the frame relay–based VPN approach?

Pros:

- Universal LAN-to-LAN network connectivity at a lower price than the private-line option.
- Higher reliability at a modest cost. The frame relay cloud is engineered by the service providers to be highly resilient. The cloud engineering provides for rerouting capability in case of a failure within the cloud. As for a site resiliency, the incremental cost of adding a backup PVC is quite minimal.
- Frame relay provides the same or better performance as compared to private-line option at 30% to 40% cost savings. Service providers also support SLAs based on the CIR value.
- Provides a high level of security, since the traffic is limited to the PVC.
- Can support multiprotocol transport based on RFC 1490.

Cons:

- Variable delays are not suitable for interactive multimedia applications with audio and video content.
- Frame relay connections are not scalable. A full mesh requires N (N − 1)/2 circuits. A partial mesh is certainly an option, but may result in suboptimal routing.
- Dynamic connections to new sites cannot be established on-demand. Adding a new site to a mesh requires installation of a new circuit and new PVCs. New circuit installs have long lead-times depending on the availability of port capacity on the provider network and availability of local loops from the LEC. New frame relay services based on SVC have made some headway in this direction.

- During congestion, frame relay–based traffic cannot be configured to discriminate between critical and noncritical traffic.

Implementation Example

Consider the IP addressing scheme shown in Table 6.1 for each router.

A sample configuration for router R1, R2 and R3 in an intranet configuration is shown below. The example assumes Cisco 4700 routers.

Router R1

ip classless ## **Allows addresses with variable subnet masks**
!
interface Serial0
description 512k frame link to MCI
encapsulation frame relay ietf ## **Conforms to RFC 1490**
interface s 0.1 point-to-point ## **The first subinterface on serial port zero**
ip address 207.80.127.1 255.255.255.252
frame relay interface-dlci 10 broadcast
!
interface s 0.2 point-to-point ## **The second subinterface on serial port zero**
ip address 207.80.127.5 255.255.255.252
frame relay interface-dlci 12 broadcast

Table 6.1
Site Router Configuration

Router	Interface		IP Address	Netmask
R1	S0.1	(DLCI 10)	207.80.127.1	255.255.255.252
	S0.2	(DLCI 12)	207.80.127.5	255.255.255.252
	E0	(HQ LAN)	207.80.153.1	255.255.255.0
R2	S0.1	(DLCI 14)	207.80.127.9	255.255.255.252
	S0.2	(DLCI 16)	207.80.127.13	255.255.255.252
	E0	(HQ LAN)	207.80.153.2	255.255.255.0
R3	S0.1	(DLCI 11)	207.80.127.2	255.255.255.252
	S0.2	(DLCI 15)	207.80.127.10	255.255.255.252
	E0	(Regional LAN)	207.80.154.1	255.255.255.0
R4	S0.1	(DLCI 13)	207.80.127.6	255.255.255.252
	S0.2	(DLCI 17)	207.80.127.14	255.255.255.252
	E0	(Vendor LAN)	204.70.143.1	255.255.255.0

```
!
router rip
network 207.80.153.0
version 2                              ## Allows classless addresses
no auto-summary
passive-interface s0.2    ## No routing updates are allowed from this interface
!
```

Router R2

```
ip classless                   ## Allows addresses with variable subnet masks
!
interface Serial0
description 512k frame link to MCI
encapsulation frame relay  ietf              ## Conforms to RFC 1490
interface s 0.1 point-to-point     ## The first subinterface on serial port zero
ip address 207.80.127.9 255.255.255.252
frame relay interface-dlci 14 broadcast
!
interface s 0.2 point-to-point    ## The second subinterface on serial port zero
ip address 207.80.127.13 255.255.255.252
frame relay interface-dlci 16 broadcast
!
router rip
network 207.80.153.0
version 2
no auto-summary
passive-interface s0.2
!
```

Router R3

```
ip classless                   ## Allows addresses with variable subnet masks
!
interface Serial0
description 512k frame link to MCI
encapsulation frame relay                     ## The global command
interface s 0.1 point-to-point     ## The first subinterface on serial port zero
ip address 207.80.127.2 255.255.255.252
frame relay interface-dlci 11 broadcast
!
interface s 0.2 point-to-point    ## The second subinterface on serial port zero
ip address 207.80.127.10 255.255.255.252
```

frame relay interface-dlci 15 broadcast
!
! route to HQ
ip route 207.80.153.0 255.255.255.0 207.80.127.1
                              ## This is the preferred path (subinterface s0.1)
! route to HQ with administrative distance
ip route 207.80.153.0 255.255.255.0 207.80.127.9 150
                                        ## This is the backup route
!

Next we will take a look at the VPN implementation using ATM.

Asynchronous Transfer Mode

The ATM-based VPN approach is much the same as the frame relay approach. The LAN-to-LAN or LAN-to-WAN connection is established through a public ATM network. The intranet and extranet sites are connected through ATM VCs. The design considerations for frame relay service and ATM service are similar in nature. Both deal with the congestion in the backbone in a similar way. Just as frame relay would mark a frame with a discard-eligible (DE) bit, a cell-relay switch will mark an ATM cell with cell-loss priority (CLP). Since ATM is a layer 2 technology, it can support multiple layer 3 protocols. There are various RFCs and ATM forum protocol specifications to accomplish this goal. Some of the specifications of interest (you can get rest of the ATM Forum specifications from http://www.atmforum.com/atmforum/specs/approved.html) are:

- RFC1483 and RFC1577 for IP over ATM;
- LANE1.0 and 2.0 specifications from ATM Forum for connecting ethernet and token ring LANs over an ATM network;
- MPOA 1.1 specifications for multiprotocol transport over ATM.

Pros:

- ATM provides a high-bandwidth connection that scales.
- It is possible to set up backbone connections based on QOS agreements.

Cons:

- ATM presents the same scaling issues as frame relay, although with PNNI, MPOA, and SVC support, these issues can be mitigated.

- If PVCs are used in a partial mesh configuration, it can result in sub-optimal routing.
- There are not many service providers offering ATM backbone service at all locations.

6.4.2   Network Layer

There are many tunneling techniques for creating network layer VPNs. Some of the tunneling specifications are listed below:

- Generic routing encapsulation (GRE);
  - RFC 1701 (GRE);
  - RFC 1702 GRE over IPv4 networks.
- IP over IP;
  - RFC 2003 IP encapsulation within IP;
  - RFC 1853 IP in IP tunneling (IPv4 over IPv4).
- RFC 1234 tunneling IPX traffic through IP networks;
- IPSec in tunnel mode.

In this section we will concentrate on the GRE and IPSec protocols.

Generic Routing Encapsulation Tunnels

GRE tunnels are created between an ingress and egress router. GRE standard (RFC 1701) is generic and can be used to encapsulate any type of protocol. If the transport protocol and the tunneled protocol is IPv4, then the RFC 1702 standard is used to deploy GRE tunnels over an IP backbone. A special version of the GRE protocol, GREv2, is used to transport PPTP protocol over the IP backbone. In this discussion we will only consider Cisco's implementation of the GRE standard.

Pros:
Cisco's implementation of the GRE standard allows one to deploy VPNs with:

- Improved performance, since the IP precedence bits from the encapsulated IP packets can be copied into the transporting IP packet. If the routers in the path implement routing considering the precedence bits, end-to-end QOS over the backbone can become a reality for VPNs.

- Additional security; Cisco IOS network layer encryption can be used with the precedence information for GRE tunnels. This can provide data privacy between the tunnel end points.

Cons:

- In most tunneling implementations, the tunnel encapsulation and decapsulation process is a process-switched activity. This means the router processor is involved with processing every packet. This naturally slows the router and increases the latency. In GRE, Cisco has introduced "fast switching" in IOS11.1 code running on Cisco2500 and 4700 series routers.
- Incorrect implementation of tunnels might result in suboptimal routing. For example, a tunnel might lend an appearance that the remote node is only one-hop away. This will affect routing protocols, especially RIP, that consider the number of hop-count while making the next-hop routing decision. In the worst case scenario, the tunnels might create wrong entries in the main routing table.
- Depending on the tunnel end points, there is a chance that some of the access control rules may be bypassed, creating security holes in the implementation.
- Some protocols may not be able to tolerate increased latency due to added packet processing in the end nodes.

Table 6.2
IP Address Assignment

Router	ISP Side IP Address on s0 Interface		CPE Side IP Address on s0 Interface	Netmask
R1	207.80.127.1		207.80.127.2	255.255.255.252
	E0	(HQ LAN)	207.80.153.1	255.255.255.0
R2	207.80.127.5		207.80.127.6	255.255.255.252
	E0	(HQ LAN)	207.80.153.2	255.255.255.0
R3	207.80.127.9		207.80.127.10	255.255.255.252
	E0	(Regional LAN)	207.80.154.1	255.255.255.0
R4	207.80.127.13		207.80.127.14	255.255.255.252
	E0	(Vendor LAN)	204.70.143.1	255.255.255.0

We can illustrate the GRE tunnel configuration on a Cisco router. Consider the network setup shown in Figure 6.5. In this example, we are implementing an extranet VPN connection between routers R2 and R4 using GRE encapsulation.

The IP addresses of the tunnel end points and the LAN interfaces are shown in Table 6.2:

Router R2

interface ethernet 0
  description HQ LAN
  ip address 207.80.153.2 *255.255.255.0*
!
interface serial 0
  description connection to ISP
  ip address 207.80.127.6 *255.255.255.252*
!
interface tunnel 0   ## **Specifies the tunnel interface**
!description Extranet Tunnel
  tunnel source serial 0
  tunnel destination 207.80.127.14
      ## **Specifies the serial interface of R4 as the remote tunnel end point**
  ip address 10.3.2.1 255.255.255.0

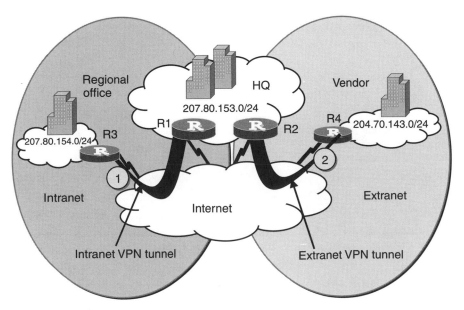

Figure 6.5 GRE tunnel configuration on Cisco router.

This address is to get IP going on the GRE tunnel

tunnel mode gre ## **Specifies tunnel mode as GRE**

Router R4

interface ethernet 0
  description Vendor LAN
  ip address 10.1.3.1 255.255.255.0
!
interface serial 0
  description connection to ISP
  ip address 207.80.127.14 255.255.255.252
  !
interface tunnel 0
!description Extranet Tunnel
  tunnel source serial 4
  tunnel destination 207.80.127.6
      ## **Specifies the serial interface of R2 as the remote tunnel end point**
  ip address 10.3.2.2 255.255.255.0
  tunnel mode gre

IP Security (IPSec)

IPSec provides a security framework for providing secure transport of IP traffic across an untrusted network. The IPSec framework consists of four main components:

1. The security-association component describes the environment under which the security services are being provided. The environment variables are unique to a session and describe the encryption, decryption, and authentication parameters specific to that session.

2. The security component consists of an authentication header (AH) and encapsulating security payload (ESP).

3. The key management component, also known as Internet key exchange (IKE), describes the protocol used to manage the participant key distribution and negotiation.

4. Algorithms are used for encryption and authentication.

Let's take look at each of these components. Once we get a basic understanding of these components, we are ready to describe how IPSec can be used to create secure VPN connections across the Internet.

Security Associations

The concept of SA is key to the working of an IPSec framework. When two IPSec-compliant devices, like two IPSec-compliant firewalls, want to set up IPSec security for the data traffic between them, they negotiate security parameters and develop an understanding of each other's capabilities and preferences. These negotiated security parameters are exchanged via the SA. Each SA is a one-way association. If the conversation between two IPSec entities is a bidirectional conversation, there is an SA for each direction of the conversation. The security parameters described in an SA are:

- The cryptographic functionality agreed upon between the two parties (for example, should the IPSec security include both the authentication header and the encapsulated security payload or just one of the two functionalities);
- Source IP address(es) or the network address;
- The cryptographic algorithms agreed upon (for example, should a combination of DES encryption be used with MD5 authentication (and integrity check) or a combination of 3DES and SHA (secure hash algorithm) be used);
- The keys used for the encryption and hash algorithms;
- Name: there are two types of names possible, userID (a fully qualified user name) or system name (like host name or a security gateway name);
- Data sensitivity level;
- Transport layer protocol (for example, IPv4 transport as TCP);
- Source and destination ports (for example, TCP or UDP ports).

Each SA is uniquely identified by a combination of security parameter index (SPI), an IP destination address, and security protocol (AH *or* ESP). Please note the "or." A single SA can define AH or ESP, but not both. If a receiving IPSec host requests both, multiple SAs are used. Each SPI is a 32-bit number that identifies a specific SA. The destination address is a unicast-, broadcast- or a multicast-group address. However, current IPSec specification defines SA management functionality for the unicast address only.

There are two types of SAs, transport mode SA and tunnel mode SA. Transport mode SA is always used between two hosts, while the tunnel mode SA is used between two security gateways or between a host and a security gateway. Various types of SA combinations are shown in Figure 6.6.

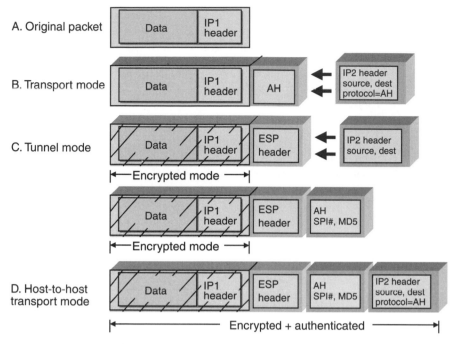

Figure 6.6 Modes of data transport in IPSec.

When a receiving node receives a SA, it looks at the SPI value and identifies the SA it is suppose to use. Once the SA has been identified, the IPSec receiver knows whether to authenticate and/or decrypt the packet and, if so, the method that should be used to authenticate the packet and the keys used to decrypt it.

Security Components

The two main security components are the authentication header and the encapsulated security payload. Except in the case of a host-to-host communication, these two components are not used together.

Authentication Header The authentication header provides authentication of the origin of the packet as well as an integrity check to make sure that the packet was not modified during the transit. The authentication is based upon a keyed one-way hash of the total packet, including the second IP header (IP2), shown in Figure 6.6(B). The hash algorithms supported are MD5 or SHA. If confidentiality is not a great concern, then AH used in the transport mode may be sufficient. In the case of an intranet or extranet VPN, confidentiality is important and, hence, the best use of IPSec is in the tunnel mode with encapsulation security payload.

Encapsulation Security Payload–Tunnel Mode   ESP adds confidentiality to the data traffic. The original packet, including the header, is encrypted as shown in Figure 6.6(C). The encryption algorithms supported are 56-bit DES, 3DES, and any other symmetric key algorithms. ESP may also provide some authentication for the upper-layer protocol. ESP may also add variable padding to hide the actual length of the original packet. A pad-length field is added to indicate the length of the padding field. The original packet plus padding fields plus padding length are all encrypted by ESP. The SPI number indicates the SA to be used and all the parameters for ESP decryption and authentication. A new transport packet is added to the ESP encapsulated packet and the packet is then sent to the destination. This mode of transport is also known as the "tunnel mode," since an IP packet is being sent inside another IP packet (though in a highly secure manner).

Note   Since the original packet is encrypted, one could transport a packet with an unregistered source and destination IP addresses.

Key Management Component (IKE Previously ISAKMP/OAKLEY)

The whole notion of IP security is based on the secure setup of authentication and encryption keys between two parties participating in a confidential dialog. The keys themselves could be negotiated once at the beginning of the dialog or negotiated many times during the dialog for added protection against a "replay" attack. In the simplest scenario, consisting of a small number of users and/or applications, the keys can be exchanged manually. This scenario assumes that the two communicating entities are known to each other. What happens when the communicating entities are not known to each other, and they are separated geographically? Add to this the requirement for multiple keys being exchanged during just one dialog and you quickly realize that the manual method of key exchange is just not scalable. You need an automated method of key exchange that provides scalability and security for the process of key exchange. As we have observed in the chapter on security, one solution to this dilemma is to use asymmetric key exchange verses the symmetric key exchange. Asymmetric key exchange has its own set of operational hurdles, for example, how do you guarantee that the public key that you have received belongs to the user who sent it? This discussion led us to the discussion of PKI and the hierarchy of CA, among other things, in Chapter 5.

IETF recognized these issues and defined ways to manage the SA's associated keys and means to exchange them securely. RFC 2408 describes Internet Security Association and Key Management Protocol (ISAKMP) while RFC 2409 describes the Internet key exchange protocol. The requirements for key exchange are described in an informational RFC describing OAKLEY protocol

(RFC 2412). We will take a brief look at these protocols and leave the details to the study of the RFCs.

ISAKMP The IPSec working group within IEFT originally defined the ISAKMP. As stated in RFC 2408, "The ISAKMP defines the procedures for authenticating a communicating peer, creation and management of security associations, key generation techniques, and threat mitigation (e.g., denial of service and replay attacks). It defines a common framework for agreeing to the format of SA attributes, and for negotiating, modifying, and deleting SAs." As we have seen in our previous discussion, each of these factors is necessary for secure communication between two entities.

ISAKMP allows two communicating entities to set up the rules of engagement in two stages. The first stage sets up the SA between the two entities and the second stage sets up the SAs for the AH and/or the ESP. The first set of SAs and associated authentication and encryption algorithms protect the second level of exchange. ISAKMP uses digital signatures for authentication. Various steps in the ISAKMP operation are shown in Table 6.3. The I and R cookies refer to the initiator and the receiver in the message exchange. Notice that the SPI value is set to zero for the SAs defined in the first stage.

OAKLEY The OAKLEY protocol provides a method for key exchange between two communicating entities to establish a shared key. The key exchange provides the IDs of communicating entities, key name, secret component of the key, and the type of algorithms used for encryption, authentication, and data integrity. The OAKLEY key exchange protocol is based on the Diffie-Hellman key exchange algorithm. OAKLEY is compatible with ISAKMP.

Table 6.3
ISAKMP Operational Steps

Operation	I-Cookie	R-Cookie	Message ID	SPI
Start ISAKMP SA negotiation	✓	0	0	0
Respond ISAKMP SA negotiation	✓	✓	0	0
Initialize other SA negotiation	✓	✓	✓	✓
Respond other SA negotiation	✓	✓	✓	✓
Other (KE ID etc.)	✓	✓	✓ or 0	NA
Security protocol like ESP and AH	NA	NA	NA	✓

IKE   IKE is a hybrid protocol that combines the features of ISAKMP and subset specifications from the OAKLEY protocol. It is the recommended standard for IPSec automated SA and key management. IKE provides:

- SA and key parameter negotiation service;
- Primary authentication for the communicating entities at the start of the negotiation;
- Management of the key exchange;
- Method for generating other keys for authentication and encryption service.

IKE operates in two phases.

Phase 1   IKE sets up a secure channel of communication between two ISAKMP peers, by setting ISAKMP SAs. The channel can be set up in two modes derived from OAKLEY: the main mode and the aggressive mode. Each mode generates an authenticated keying material from a key exchange based on the Diffie-Hellman algorithm. The ISAKMP setup process is similar to the one identified in Table 6.3. A single phase 1 negotiation can be used to set up multiple phase 2 negotiations. The main mode operation is slower than the aggressive mode operation. This is due to the fact that the main mode operation requires identity protection. The ISAKMP SA negotiates the algorithms for encryption (DES, 3DES), authentication (digital signatures, preshared keys), and data integrity (hash algorithm, MD5, SHA).

Phase 2   In phase 2, IKE is used to negotiate and set up SAs for other services like AH and ESP. The setup is accomplished using the quick mode of operation. All information exchange in the quick mode is protected by the ISAKMP SA. Quick mode also provides exchange of nonces (pseudo-random function for protection against replay attack). The nonces are used to generate fresh keys during the secure exchange.

Pros of IPSec:

- IPSec allows you to build VPNs that are standards-based and interoperable with any IPSec-compliant devices like firewalls and end-nodes.
- IPSec allows you to build a scalable solution that incorporates many layers of security.
- The keying mechanisms specified in IPSec standards allows for multiple rekeying operations, improving the security of data exchange and preventing any replay attacks.

Cons of IPSec:

- IPSec implementation is limited to IP-only environment.
- IPSec introduces a penalty on the operation of the end-host and the security gateway implementing IPSec.
- The performance penalty also translates into increased latency and potential reduction in throughput.
- Use of IPSec tunneling can result in reduction in bandwidth utilization due to the increased size of the packet, especially in the tunneling mode. In the transport mode the AH header adds to the length of a packet. This could result in potential packet fragmentations, adding to the performance penalty.
- PKI infrastructure is a must for IPSec with multiple dynamically generated keys.

Various IPSec solutions are now available in the market with support for IKE. Some of the leading vendors are TimeStep, RedCreek, RADGUARD, VPNet, Checkpoint, and Cisco. The first four vendors provide a hardware-based solution that can minimize the impact of IPSec-related computations. Even the software-based vendors have realized that they need to provide a hardware-based solution and are investigating means to meet this market demand.

6.4.3 Session Layer

Two prominent technologies for setting up session layer VPN networks are:

- SOCKSv5;
- SSL 3.0.

SOCKSv5 Plus SSL

SOCKS provides session-level proxy, also known as circuit-level proxy. A circuit-level proxy establishes a proxy connection between an internal and external entity. It intercepts the information between two entities and relays it based on the proxy setup. An internal entity could be a client side of the application connecting to an external server. Unlike application proxy, which requires a proxy service for each application, SOCKS does not care what applications are being proxied through it. This is because unlike application proxy, SOCKS operates at the session layer.

In the past (SOCKSv4), the client side of the application had to be "sockified" (i.e., the client application had to be recompiled with SOCKS libraries, assuming that the source code was available). Now with SOCKSv5 and the software products from NEC and Aventail, the shared libraries (for UNIX) and link libraries (for Windows) make it possible to provide the sockification service at the run time. Aventail client software uses Microsoft's Layered Service Provider (LSP) architecture in the Winsock 2.0 environment.

Pros:

- Because SOCKS is providing proxy service for internal clients, it can hide the nature of the client platform and the internal network from external entities.
- Since SOCKS is application independent, it provides a flexible solution that can be deployed under any application requirements and does not need to be changed when new applications are added on the network.
- All traffic going through the circuit proxy can be logged and analyzed in real-time or after the fact.
- Provides support for authenticated client/server communication. Different types of authentication methods are supported and are negotiated during the connection setup between the application client and the SOCKS server. Examples of supported authentication methods are: regular password, CHAP, S/Key, digital certificates, and token cards like SecurID card from Security Dynamics.
- SOCKSv5 provides support for standards-based GSS-API (general security services, RFC 2078) and negotiates the message integrity and confidentiality algorithms. The encryption algorithms supported are DES, 3DES, and RC4.
- SOCKS server is available on a number of different OS platforms.
- SOCKSv5 supports UDP traffic. Since UDP traffic is a stateless protocol, this is a great advantage.
- SOCKSv5 is firewall independent and allows for authenticated firewall traversal.

Cons:

- SOCKS implementation needs a client-side software that can communicate with the SOCKS proxy server. When SOCKS client software detects that the traffic is destined to the network protected by SOCKS proxy, it directs all traffic to the SOCKS proxy server.

- Since SOCKS is unaware of the application-level traffic, it cannot provide highly granular application-level control in the proxy server.
- Only IP protocol is supported.
- Since the proxy server manages all the communication, it can become a single point-of-failure.

SOCKSv5, when used in conjunction with secure socket layer (SSL) allows a network-level encryption and certificate-based authentication. The beauty of this solution is that SSL has been incorporated in many client applications including browsers. All the firewalls allow use of SSL. This provides a low-cost universal-access solution for accessing intranet or an extranet application server over the Internet. SOCKSv5 has been gaining steady acceptance in the Internet community. Vendors like Aventail and NEC are providing SOCKSv5-based intranet and extranet access solutions today.

6.4.4 Application Layer VPN Solution

An application layer solution combines the application-level proxy, SSL, and PKI solution to provide authenticated and encrypted VPN solutions. By virtue of the SSL 3.0 implementation, you have a choice of RC2, RC4, DES, or 3DES encryption algorithms. The authentication server is an X.500 server or an LDAP server.

Other application layer encryption solutions include proprietary protocols. Some of the examples of application layer security are S/MIME and PGP. Raptor and Guantlet are examples of application layer proxy firewalls.

Pros:

- Available with a stable code that is easy to implement;
- Application layer of granularity is available for implementing security controls.

Cons:

- New application layer proxies need to be written whenever new applications are added behind the proxy server;
- Potential single point-of-failure;
- Due to proxy involvement, the end-to-end performance may be affected.

6.4.5 Dial-to-LAN VPDN Connection

The dial-to-LAN VPN solutions provides dial-in access to a corporate intranet or an extranet. This type of VPN solution is ideal for road warriors and telecommuters. The requirements for these two types of users differ slightly. In the case of road warriors, the requirements are:

- Universal connectivity;
- Ease of use;
- Willingness to give up on speed to get universal access;
- Problems resolved by help desk immediately;
- An 800 number (800 numbers are available in North America only) or a local dial number that somehow connects them to the remote LAN via VPN.

In the case of telecommuters, the requirements are:

- A high-speed connection to the LAN in order to work at the same level of productivity as in an office environment;
- A local number, since the users are expected to be on-line for long periods of time;
- Ease of use;
- Access to the right set of equipment, for example a desktop computer, printer, and a second telephone line for conference calls and fax.

In order to provide secure dial-in access to the LAN, there are two options. One is to build a dial infrastructure in-house and ask all users to dial-in using local, long distance, or an 800 number. The second option is to use the wonderful Internet infrastructure to set up a universal-access infrastructure using VPN solutions. The first option is obviously simple to set up, since all the management of the dial infrastructure is done centrally. This advantage disappears very quickly if you have a large national or international organization and need to provide central and regional dial infrastructure. Add to this the cost of administrative personnel, training, and the cost of 800 and long-distance calls. The cost increases almost exponentially.

The alternative is a VPN-based remote access solution offering universal connectivity at a lower price but with an extra level of complexity. New solutions from vendors are always striving to eliminate this extra complexity by making the solutions easy to use and administer. In this section we are going to discuss the data link layer solution to the dial-VPN requirement. Other solutions based on IPSec are left for the case study.

There are two types of layer 2 protocol implementations that are suitable for dial-up intranet or an extranet VPN solution:

1. Point-to-point tunneling protocol (PPTP);
2. Cisco's layer 2 tunneling protocol (L2TP).

Both protocols address the following dial-VPN solution requirements:

- Strong encryption;
- Strong and auditable authentication;
- Strong key management;
- Address management for dial-in node;
- Multiprotocol support for dial-in connection since the LAN the user is dialing into may be based on non-IP protocols like IPX.

Since both PPTP and L2TP operate at layer 2, they are oblivious to any protocols being used on layer 3 and above. Both PPTP and L2TP encrypt the data traffic and send it encapsulated within a transport protocol. While PPTP is constrained to IP transport, L2TP can operate over any transport medium that can support point-to-point delivery of a PDU (protocol data unit) such as an IP, X.25, frame relay, or an ATM network. Both PPTP and L2TP protocols are based on the operation of PPP protocol. The choice of the protocol depends on who has the ultimate control over the VPN connection, the user, or the Internet service provider. We will discuss the selection criteria after we have explored each protocol in detail.

Point-to-Point Tunneling Protocol

PPTP protocol is a layer 2 protocol that encapsulates the PPP frames in IP packets for transmission over any IP backbone. The IP backbone could be a private IP backbone or the public IP backbone like the Internet. The PPP packets are encapsulated in IP using the GRE protocol. The PPTP protocol is currently an IETF draft (draft-ietf-pppext-pptp-08.txt) and is jointly produced by Microsoft, Ascend, 3Com, and other players in the remote access market. The biggest advantage of PPTP is that it is free and shipped pre-built into the Windows NT server, NT Workstation, Windows 98 desktop and available as an add-on module to Windows 95 desktop as well. With the coverage for 85% of the worldwide-desktop market and almost 50% of the server market, this is a very attractive solution to deploy. Let's take a detailed look at the inner workings of PPTP protocol, its components and configuration in Windows environment.

Inside PPTP

PPTP protocol specification consists of two main components:

- The control channel;
- The data channel.

The control channel is used to set up data channel parameters like encapsulation, encryption, and payload compression. The control connection is responsible for the establishment, management, and release of the data channel connection, also known as the "tunneled connection." The control channel is a standard TCP connection.

Once the control connection has been established between the PPTP client and the PPTP server, it is now time to set up the data channel connection. In a data channel connection, PPTP protocol encapsulates IP, IPX, or NETBUI PDUs using PPP protocol. The PPP packets are then further encapsulated using an extended GRE (GREv2) encapsulation scheme and transported in an IP packet. The PPTP-based VPN can be implemented in two ways. One way is to initiate the PPTP session from a client to the remote PPTP server. Another way is to let the NAS operated by an ISP initiate the PPTP session on the user's behalf to the remote PPTP server. There are pros and cons of each approach. The operation of PPTP in each case is shown in Figures 6.7 and 6.8.

In the case of a client-initiated PPTP session, the IP, IPX, or NETBUI datagram is encapsulated in a PPP-protocol unit and a PPP connection is set up with the ISP NAS. After the successful PPP(1) setup, a second dial-up call is made to the remote NT server, over the existing PPP connection. The new PPP(2) session is encapsulated using GREv2 PPTP encapsulation and then sent over the Internet to the remote router connected to the destination PPTP server. The remote router strips away the GREv2 encapsulation and forwards the PPP(2) encapsulated frame to the PPTP server. The PPTP server strips away the PPP(2) encapsulation and processes the original IP, IPX, or NETBUI datagram. The PPTP client logs into the PPTP server using the CHAP, MS-CHAP, or PAP authentication. The inner PPP(2) frame could be compressed and encrypted.

In the case of an NAS-initiated PPTP connection, first the control connection is established between the NAS and the PPTP server. In this scenario, the NAS acts as a PPTP client. When the NAS receives a request destined to the PPTP server, it sets up a PPTP control connection with the PPTP server. The

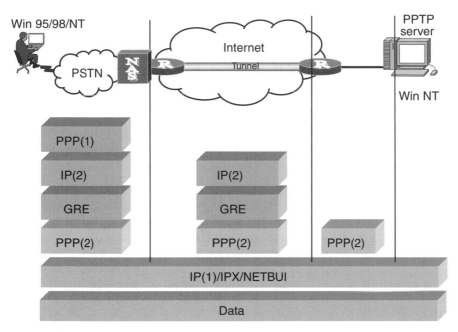

Figure 6.7 Client-initiated PPTP session.

received PPP(1) packet is decapsulated and the original packet is re-encapsulated into a PPP(2) frame which in turn is encapsulated into an IP packet using GREv2 encapsulation. This IP packet is then sent to the destination router over the Internet. The destination router removes the GRE encapsulations and passes the PPP frame to the PPTP server.

The NAS-initiated PPTP option, when available, is preferable to the client-initiated approach. This is because the NAS-based approach simplifies the installation and configuration of the client desktop or laptop, thus reducing the help desk calls and the cost of providing IT support. The PPTP support is now being added to the Ascend, 3COM, and Cisco network access servers though most ISPs are shying away from providing PPTP support for fear of added support cost and potential liability considerations for providing a VPN solution.

Some of the installation screens on a Windows NT server and Windows 98 are shown below:

Windows 98 Configuration:

Step 1: Add VPN adapter under Control Panel → Network → Add → Adapter → Microsoft → Virtual Network Adapter (See Figure 6.9).

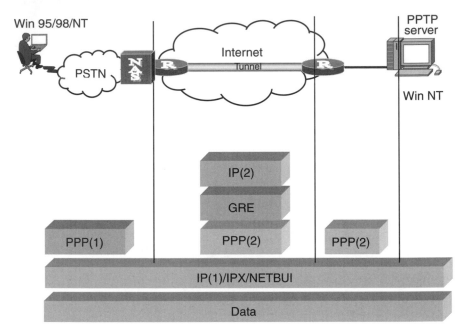

Figure 6.8 NAS-initiated PPTP session.

Step 2: Now add a new dial-up connection under Dial-up Networking and name it "PPTP_Connection" (You can of course choose any name!). Next, specify the address of the remote VPN server (see Figure 6.10(a) and (b)).

Step 3: The Next button will complete the process of creating a "PPTP_Connection." An icon will appear under the Dial-up Networking window. If you right-click on the icon and select Properties, you will see the screen shown in Figure 6.11.

Step 4: On the NT-server side, select Control Panel → Networking → Protocol → Add → PPTP network protocol.

Step 5: Next, select the maximum number of simultaneous PPTP sessions that you plan to support on this server (see Figure 6.12).

Step 6: Once the network protocol setup has been completed you will be guided through the RAS setup dialogs to configure new RAS devices derived from Step 5. These VPN ports are configured for dial-in connection only, though they can be configured for a dial-out connection as well. The VPN ports can be configured just as any other RAS port.

Here is a summary of the pros and cons of the PPTP-based VPN setup.

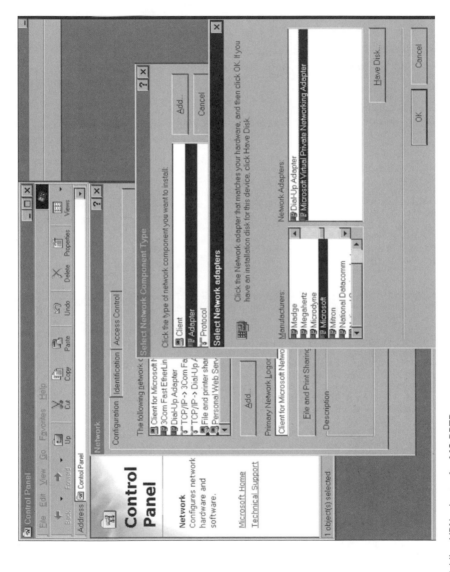

Figure 6.9 Adding VPN adapter for MS PPTP.

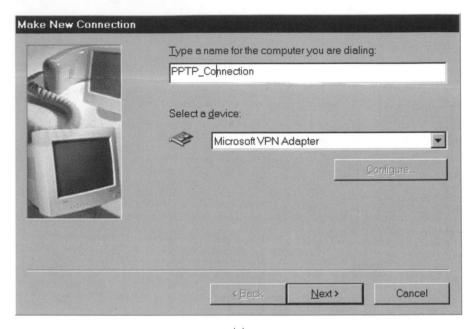

(a)

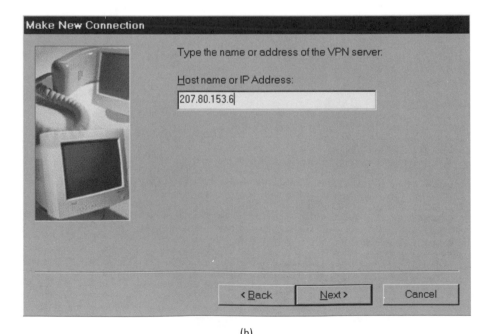

(b)

Figure 6.10 (a, b) PPTP connection setup on client.

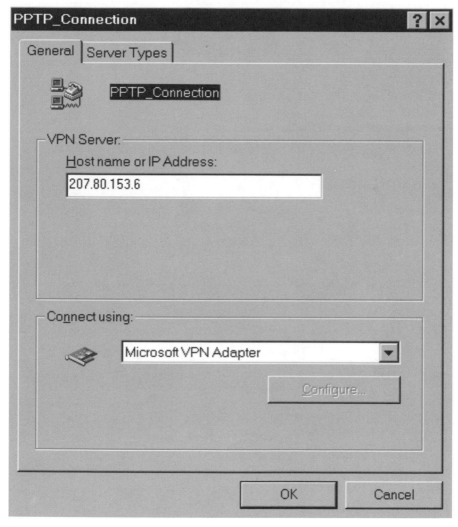

Figure 6.11 PPTP connection properties on client-side configuration.

Pros:

- Supported on the largest installation of desktop operating systems.
- It is free.
- Provides multiprotocol support.
- Allows use of existing LAN network addresses. Since the PPP session with the PPTP server is tunneled while on the Internet, a PPTP server

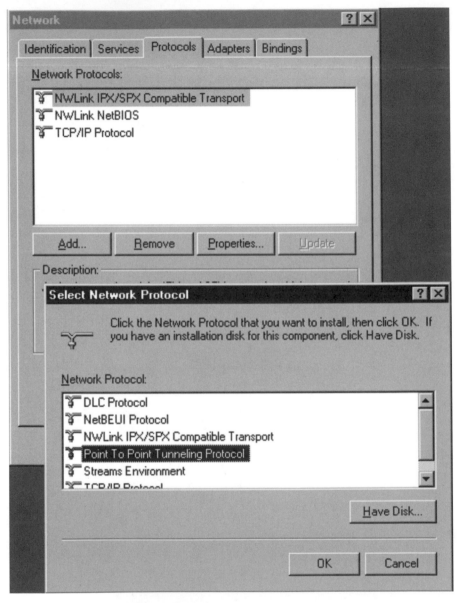

Figure 6.12 Server-side PPTP configuration.

could assign any IP address to the PPTP client, including private IP addresses.

- Does not need any change to the ISP connection.
- PPTP protocol standards have been published.

Cons:

- Scalability of the solution is still questionable. When a large number of PPTP tunnels are being terminated on an NT server, the processing power of NT can be a limiting factor.
- PPTP requires an IP infrastructure. Since the Internet and most of the large enterprises using NT are already IP based, this limitation may not have any significant impact on deployment of PPTP.
- The control channel of PPTP is not authenticated. A malicious user could cause a termination of PPTP sessions managed by the control channel.
- The MS-CHAP NT hash response is vulnerable to a dictionary attack.

There are other vulnerabilities that get reported on NT Bugtrack (NTSECURITY@LISTSERV.NTBUGTRAQ.COM) and other NT-related security (NTSD@LISTSERV.NTSECURITY.NET) mailing lists. It is a must for an IT administrator to keep on top of the reported issues and the Microsoft fixes. Some "hot fixes" are so critical that you should not wait for the next major release of the Microsoft service pack.

Layer 2 Tunneling Protocol

Layer 2 tunneling protocol is another method of tunneling layer 2 frames over a packet-switched backbone like the Internet, frame relay, ATM, or X.25. L2TP combines the best of both features from the PPTP specifications and Cisco's L2F (layer 2 forwarding) specification. Like PPTP, L2TP is designed to be used by end users or by service providers. In a service provider–supported L2TP VPDN, L2TP operates between a service provider LAC (layer 2 access concentrator, also known as PAC in PPTP specification and generally known as NAS) and a corporate LNS (layer 2 network server, also known as PNS in PPTP specification). The idea is to benefit from a local termination of a layer 2 call and tunnel the call to the remote NAS over a packet-switched backbone, thus benefiting from the cost savings on a long distance call and having the ability to maintain the control over the security of incoming connection.

The latest draft of the L2TP specification is draft-ietf-pppext-l2tp-14. Microsoft is planning to support L2TP in the next release of Windows NT (Windows 2000) operating system. L2TP is currently supported by Cisco on its routers and NASs.

Operation of L2TP is shown in Figure 6.13.

Step 1: User PPP request is received at the LAC.

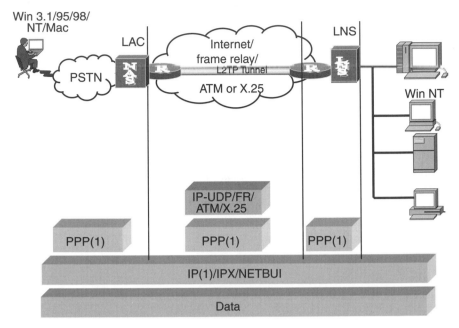

Figure 6.13 NAS-initiated L2TP tunnel operation.

Step 2: LAC performs the LCP negotiation including the PPP authentication. The authentication can be programmed to be as simple as the userID@domain.

Step 3: Based on the destination domain, LAC initiates a control connection to the domain LNS server to set up the L2TP tunnel.

Step 4: Once the tunnel has been established, the PPP packets received are encapsulated into an L2TP tunnel (as per the setup parameters negotiated by the control connection, including encryption and header compression) and sent to the destination LNS.

Step 5: LNS decapsulates the packets and presents it to the remote PPP server based on the security policy configured on the LNS.

Step 6: The end user negotiates a PPP session with the remote PPP server. The PPP server can assign an IP address from the local LAN address space, thus connecting the user seamlessly to the LAN over a wide-area network.

A comparison between PPTP, L2F, and L2TP protocols and their operational requirement are shown in Table 6.4.

Table 6.4
Comparison of PPTP, L2F, and L2TP

Specifications	PPTP	L2F	L2TP
Wide area network requirement	IP only	IP, FR, ATM, or X.25	IP, FR, ATM, or X.25
User initiated tunnel	Yes	No	Yes
ISP NAS initiated tunnel	Yes	Yes	Yes
Authentication protocol supported	CHAP, PAP, MS-CHAP	CHAP, PAP, MS-CHAP	CHAP, PAP, MS-CHAP
Multi-link PPP support	Yes	Yes	Yes
Control channel authentication	None	Mandatory	Optional
Flow control	Yes	No	Yes
Number of tunnels supported between endpoints	Single	Multiple	Multiple
Header compression supported	No	Yes	Yes

6.5 Conclusion

In this chapter we have looked at various virtual private network technologies available for implementing secure VPN connections between two LAN locations and between a dial-in user and a LAN location. The technologies discussed apply to both intranet and extranet implementations. The LAN-to-LAN VPN technology could be used to connect multiple intranet locations over a public packet-switched network like the Internet, frame relay, ATM, or X.25. The LAN-to-LAN (WAN) connection could also be used for connecting an extranet partner to the extranet servers located on the intranet. The dial-in VPN technology could be used to provide ubiquitous remote access to roaming employees, telecommuting users, or partners. The VPN technology like PPTP, L2TP, SSL, and IPSec allows us to maintain the privacy and integrity of data on a public network without having to deploy an expensive private-line-based data network. The selection of which technology is best suited for an enterprise depends on the needs of the organization, the cost of deployment, and the security approach within the organization.

7

Case Studies

After considering the various components of intranet and extranet, it is time to build one, beginning with an intranet of a fictitious company, NetInfo Travel Services, Inc. (NIT). NIT is based in New York City and has a workforce numbering 150. To illustrate the design and implementation aspects of building a growing intranet and extranet, we are going to follow the growth of NIT from a one-office operation to an international company having four offices in the United States and several offices in Europe. We will build NIT's intranet in phases, in-sync with the growth of the company. NIT's intranet and extranet implementations will be covered in the following case studies:

- *Case Study I:* Intranet in a company having a single office location.
- *Case Study II:* Intranet in a company having multiple office locations spread over a wide geographical area.
- *Case Study III:* Intranet in a company with a legacy X.25 connection to its operations in Europe.
- *Case Study IV:* Intranet in a company with access to an IBM mainframe system using SNA over frame relay.
- *Case Study V:* Intranet connectivity in a company using Internet-based VPN.
- *Case Study VI:* Remote access to an intranet using Internet-based VPN.
- *Case Study VII:* Extranet access via VPN over the Internet.

Each case study strives to illustrate design and analysis of intranet and/or extranet solutions with a unique approach to implementation based on a multi-vendor approach.

While we designed NIT as a travel services company for this case study, the principles explained here apply to any company in any industry planning to implement or expand its corporate intranet and extranet.

7.1 Case Study I: Intranet in a Company Having a Single Office Location

7.1.1   Case Study Objective

If you are a small to medium-sized business with a single office location, Case Study I guides you through the process of setting up your corporate intranet. We consider network planning, architecture, and security aspects in implementing the infrastructure for this intranet. We also provide configuration details on various network and security components in the architecture. One of the highlights of this case study is in designing and implementing a secure dial-in access to the corporate intranet. This case study does not include details of the intranet applications.

7.1.2   Case Study Background and Requirements

NIT is a travel services company located in New York City. The company is grouped into sales, customer service, human resources, and finance departments. The current mode of communication between the departments is based on shared file servers on the network. NIT's vision of corporate intranet includes implementation of a secure Web-based information sharing system that provides secure access to company employees on the LAN as well as employees who dial into the intranet. NIT's ROI analysis has indicated high-efficiency gains with such an intranet.

NIT's current network architecture uses TCP/IP protocol and interconnects multiple 10BaseT ethernet hubs running TCP/IP. The company would like to overhaul its network infrastructure, making it more efficient and secure. Additional desired features of this intranet would be support for secure dial-in access to the intranet for mobile and telecommuting employees. Employees and administrators would use the secure dial-in access for accessing the intranet and other LAN resources remotely. The remote intranet access is an important requirement for the company's sales force automation program. NIT is also planning to connect the corporation to the Internet in the near future and, hence, this requirement should be considered during the network and security design of the intranet.

Architecture

The company can meet its requirements of higher security and better network efficiency by isolating the interdepartmental traffic. Thus, someone in customer support cannot access the systems in the finance department or the human resources department. The various departments in the company can be grouped together as follows:

- Front office: sales and customer services;
- Back office: human resources and finance.

In implementing the traffic localization and isolation strategy, NIT must create a hierarchical network architecture with separate subnets (mini-LANs) for each department or functional group. This can be accomplished by splitting big LANs into multiple, smaller LANs. The traffic isolation can also be accomplished using simple bridges, but this solution does not provide additional security. An alternative that accomplishes both traffic isolation and increased security is based on the use of a router that interconnects the departmental subnets. Each departmental subnet is on a separate ethernet interface on the router. (Another solution that accomplishes the same goals is using a combination of ethernet switch and a router.) NIT can now implement various security policies by using the ACLs on the router ethernet interfaces (refer to discussions on ACLs and security policy in Chapter 5). Figure 7.1 provides a look at network topology before and after the subnetworking of the single corporate LAN and depicts the creation of a hierarchical network from a single flat network. The "cloud" represents a group of workstations and resources, such as printers, associated within a department.

Network Components

Let us take a look at the network components in the hierarchical network architecture. The two key components are the router and 10BaseT ethernet hubs for each subnetwork.

With the introduction of new subnetworks, the IP addressing must be changed to reflect the new subnets. We are going to take a detailed look at each one of these components.

Router

What is the configuration of such a router? We can specify the router after we have identified the following design parameters for the network:

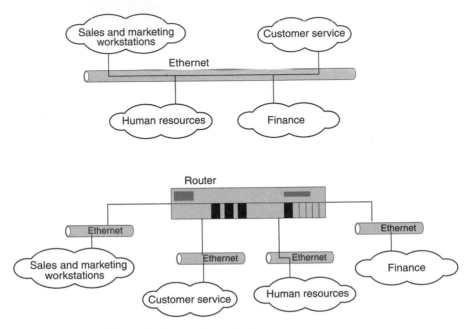

Figure 7.1 Flat and hierarchical network topologies.

- Number and type of LANs. *Example:* The type of LAN could be ethernet, FDDI, ATM, or token ring.
- The LAN speed. *Example:* 10/100/1,000 Mbps in the case of ethernet while in the case of a token ring this could be 4 or 16 Mbps. One also needs to find out if the data transmission supported is full-duplex or half-duplex.
- Type of WAN interface required and its transmission speed for future Internet connection. *Example:* Serial T1.
- Protocol support required. *Example:* IP, IPX, AppleTalk.
- Routing protocol support for interior and exterior gateway routing protocol. *Example:* RIP2 (routing information protocol version 2), OSPF, IGP for interior, and BGP for exterior.
- Size of the routing table determines the configuration of the router memory and the desired router processing power.
- Security design drives the features that must be supported by a router.

Let's consider each one of these parameters. As per our previous discussion, the various departments in NIT have been grouped together into two major groups, the front-office group and the back-office group. We must create a separate LAN for each of these groups. An additional LAN needs to be created for consolidating common application servers like the intranet Web server and the DNS server. Thus, in this intranet, the router needs to support at least three ethernet LANs plus accommodate for future growth in the number of LANs. The LAN technology of choice is 10BaseT ethernet.

To support future Internet connections, the router has to have at least one serial interface. Most of the standard serial interfaces on routers can support speeds from DS0 (64 Kbps) up to DS1 or T1 (1.544 Mbps). For traffic speeds higher than a T1, the router needs multiple T1 interfaces or a T3 (45 Mbps) interface. NIT foresees its initial Internet connection to be a fractional T1 (256 Kbps) growing to a full T1 within a year.

Routing table entries in a LAN of this size are quite small. Depending on the number of connections to the Internet, NIT can explore two options.

1. To set up a default static route to its ISP;
2. To set up a dynamic route using BGP protocol.

Option 1 Pros:

- The router has a simple configuration and operational simplicity.
- The memory and processing requirements on the router are small.

This option is generally the first option of choice for many ISPs and corporations during the initial phases of their Internet connection due to the low cost and operational simplicity.

Option 1 Cons:

- It introduces a single point-of-failure with a single link to the Internet.

The solution is to set up multiple connections to the Internet with each connection going to a different ISP. In this case, you can use the floating static route configuration or the dynamic configuration with full routes to the Internet.

Note Floating static routes are created by setting the metric associated with a static route higher than that for a route learned via dynamic routing protocol. Thus, the route learned via the dynamic routing protocol takes precedence over the static route. If for any reason the router stops learning the route via the dynamic routing protocol, it will resort to the static route as the route of last resort.

Option 2 Pros:

- The full routes allow NIT to participate in intelligent routing and load balance their traffic to the Internet.

Option 2 Cons:

- The disadvantage of taking the full routes (i.e., all the routing prefixes for various network numbers on the Internet) is that the routing table could grow to almost 45,000 entries, thus increasing the requirements on the memory and processing power of the router.

Recommendation

Based on the given design considerations, it is recommended that NIT should purchase a router with the following specifications:

1. Minimum four 10BaseT ethernet interfaces;
2. Minimum one serial interface;
3. 100 MHz route processor and 32 MB RAM;
4. Support for IP routing protocol.

Examples of some routers matching these specifications are: (1) Cisco 4700 series router (see Figure 7.2) running Cisco IOS 11.1, and (2) Motorola MP6520 router. As Cisco is a leading vendor in the router market, we are going to use the Cisco 4700 router in this case study.

Hubs

There are quite a few manufacturers of 10BaseT ethernet hubs (see Figure 7.3). Some prominent manufacturers are 3Com and Bay Networks.

Secure Dial-in for Remote LAN Access

Requirements

Let's consider the requirements for secure dial-in access to the LAN. The remote dial-in access provides access to LAN applications like e-mail, databases, UNIX and NT file servers, and Intranet Web servers. Since the protocol of choice for NIT is TCP/IP, the dial-in access server needs to support IP protocol routing. The dial-in users use Windows-based clients. At first, NIT gives plans to provide remote access only to its senior executives plus the network and Web administrators. The total number of remote access users is 25 at first and later goes up to 100. NIT needs a scalable solution that can provide the secu-

Figure 7.2 Cisco 4700 router.

rity and ease of administration. All the incoming connections use the analog dial-up connection and must be authenticated. The authentication mechanism is based on the RADIUS or TACACS+ authentication. For additional security, users are issued secure ID cards for token-based authentication.

Recommendations

Based on the requirements outlined, it is recommended that NIT purchase a device such as a Cisco 3640 or BayNetworks' Remote Annex 4000, with 32-port asynchronous network module and a minimum of one ethernet interface (see Figure 7.4).

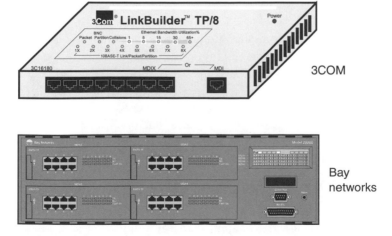

Figure 7.3 10BaseT ethernet hubs.

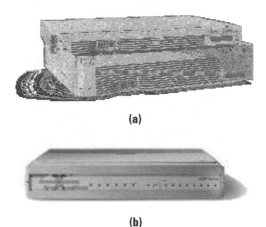

(a)

(b)

Figure 7.4   (a) Cisco 3640 and (b) BayNetworks' Remote Annex 4000 remote access devices.

Both solutions need external modems that attach to the asynchronous serial interfaces. The Cisco 3640 supports various authentication schemes including PAP, CHAP, RADIUS, or TACACS+ and will be used in this case study. The overall network architecture based on the discussed design considerations is shown in Figure 7.5.

Implementation Details

In this section we will cover the configuration details of various network components like the router and the remote access server. Please refer to the architecture diagram shown in Figure 7.5 for the rest of the discussion in this section.

Configuration of the Cisco 4700 Router

Security Considerations

To implement a consistent security policy across the enterprise, the security policy must be known to all the employees and should be published on the intranet Web site as well as distributed by other means like e-mail and booklets. (Please refer to the discussion on security policy in Chapter 5.) As for implementing these policies at the network level, we need to define the access policy at the subnet level on each ethernet interface. Such policies can be implemented using the access control rules defined in the access control lists or ACLs on Cisco routers. Table 7.1 illustrates the configurations on the Cisco 4700 router. Please refer to the configuration statements starting with keyword "access-list."

Let's take a close look at the access list 131.

1. The access list 131 is applied to the production and administration subnet on ethernet interface e0. The first line in ACL 131 allows all

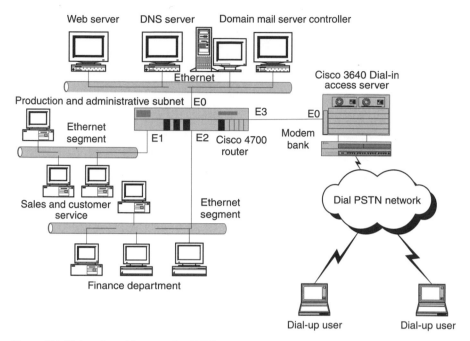

Figure 7.5 Network architecture for NIT intranet.

the connections that have been initiated from within this subnet to be completed by allowing the responses to pass through the interface.

2. The second line is an antispoofing filter that denies any incoming packets with a source address the same as this subnet. This prevents spoofing of internal address space from an external source. These two security policies have been uniformly applied to all ethernet subnets onto the router.

3. The access list 131 permits connection requests to the TCP ports 80, 443, 53, 25, and 110. Connections to the DNS server on the UDP port 53 are also permitted.

4. The last line in the access list permits the ICMP request packets through the interface. All other interfaces on the router deny these types of packets.

Configuration Details of the Cisco 3640 Remote Access Server

The Cisco 3640 uses the same Cisco IOS operating system as the Cisco 4700 series routers. As per the requirements, Cisco 3640 has been configured with one 32-port asynchronous network module (NM-32A) and one four-port ethernet

Table 7.1
Cisco 4700 Router Configuration

```
Router#wri t
Building configuration...

Current configuration:
!
! Last configuration change at 10:45:17 EDT Wed Dec 24 1997 by xxx
! NVRAM config last updated at 17:20:20 EDT Mon Dec 22 1997 by xxxx
!
version 11.1
no service finger
no service pad
service timestamps debug datetime msec localtime show-timezone
service timestamps log datetime msec localtime show-timezone
service password-encryption
service compress-config
no service udp-small-servers
no service tcp-small-servers
!
hostname router
!
clock timezone EDT-5
clock summer-time EDT recurring
boot system flash
boot system rom
enable secret 5 yyyy
enable password 7 ye4rftz3e2w
!
ip subnet-zero
no ip source-route
!
interface Serial0
description For future 512k, link to the Internet
no ip address              ## The interface is not being used in this case study
shutdown
!
interface Serial1                     ## The interface is not being used
```

Table 7.1 *(continued)*

```
 no ip address
 shutdown
!
interface Ethernet0                              ## This sets the ethernet interface
 description Production Server subnet
 ip address 207.80.127.33 255.255.255.224  ## Sets the interface IP address with the mask
 ip broadcast-address 207.80.127.63       ## The mask is a VLSM creating 32 host subnet
 ip access-group 131 out                       ## Sets the ACL on this interface
 no ip redirects
 no ip proxy-arp
 no keepalive
!
interface Ethernet1
description Sales and Customer service subnet
 ip address 207.80.127.65 255.255.255.224  ## Sets the interface IP address with the mask
 ip broadcast-address 207.80.127.95       ## The mask is a VLSM creating 32 host subnet
 ip access-group 132 out                       ## Sets the ACL on this interface
 no ip redirects
 no ip proxy-arp
 no keepalive
!
interface Ethernet2
description Finance Department subnet
 ip address 207.80.127.97 255.255.255.224  ## Sets the interface IP address with the mask
 ip broadcast-address 207.80.127.127      ## The mask is a VLSM creating 32 host subnet
 ip access-group 133 out                       ## Sets the ACL on this interface
 no ip redirects
 no ip proxy-arp
 no keepalive
!
interface Ethernet3
description Remote Access subnet
 ip address 207.80.127.129 255.255.255.192## Sets the interface IP address with the mask
 ip broadcast-address 207.80.127.191      ## The mask is a VLSM creating 64 host subnet
 ip access-group 134 out                       ## Sets the ACL on this interface
 no ip redirects
```

Table 7.1 *(continued)*

```
  no ip proxy-arp
  no keepalive
!
access-list 131 permit tcp any any established ## Passes all the established connections
access-list 131 deny ip 204.80.127.32 0.0.0.31 204.80.127.32 0.0.0.31  ## Anti-spoofing Filter
access-list 131 permit tcp any any eq www
                                   ## Allows all the web connections on Port 80
access-list 131 permit tcp any any eq 443  ## Allows all the web connections on Port 443
access-list 131 permit tcp any any eq 53
                                  ## Allows all the DNS connections on TCP Port 53
access-list 131 permit udp any any eq 53
                                  ## Allows all the DNS connections on UDP Port 53
access-list 131 permit tcp any any eq 25  ## Allows all the SMTP connections on Port 25
access-list 131 permit tcp any any eq 110
                                 ## Allows all the POP3 connections on Port 110
access-list 131 permit icmp any any    ## Allows all the ICMP connections to the subnet
access-list 132 permit tcp any any established
                                ## Passes all the established connections
access-list 132 deny ip 204.80.127.64 0.0.0.31 204.80.127.64 0.0.0.31  ## Anti-spoofing Filter
access-list 132 permit tcp 204.80.127.0 0.0.0.255 any eq telnet
              ## Allows all the telnet connections from all the address space in NIT
access-list 132 deny icmp any any
access-list 133 permit tcp any any established
                                ## Passes all the established connections
access-list 133 deny ip 204.80.127.96 0.0.0.31 204.80.127.96 0.0.0.31  ## Anti-spoofing Filter
access-list 133 deny icmp any any
access-list 134 permit tcp any any established
                                ## Passes all the established connections
access-list 134 deny ip 204.80.127.128 0.0.0.31 204.80.127.128 0.0.0.31
                                                ## Anti-spoofing Filter
access-list 134 deny icmp any any
!
  snmp-server community goodsnmp
!
line con 0
  exec-timeout 15 0
```

Table 7.1 *(continued)*

password 7 yzcv45rfs6fgt$#eth%	**## Encrypted password**
line vty 0	
access-class 109 in	
access-class 109 out	
exec-timeout 15 0	
password 7 6dtysb%3s8$frdtsr	
line vty 1 4	
access-class 109 in	
access-class 109 out	
exec-timeout 15 0	
password 7 5thrs^srfv78fvsdt5t	
!	
end	

network module (NM-4E). We will consider the configuration details for each module in the following examples.

The Four-Port Ethernet Network Module The following steps outline the configuration of the ethernet interface on the Cisco 3640 (see Figure 7.6).

1. Connect to the console port on the Cisco 3640 using the console cable.

2. Make sure the Cisco 3640 is running Cisco IOS 11.2(6)P version. This is a prerequisite for the four-port ethernet network module.

3. Login to Cisco 3640 and enter the enable mode, "Access#". "Access" is the name assigned to Cisco 3640. We will use this name in the rest of this discussion.

For configuration commands see Figure 7.6.

The 32-Port Asynchronous Network Module The following steps outline the configuration of the asynchronous interface on the Cisco 3640.

This network module has four groupings of eight ports each. The group numbering starts from the bottom-right corner as shown in Figure 7.7. The interface number for a port depends on the position of the network module in various slots in the Cisco 3640. In our case we will be inserting this module in slot 1 since the four-port ethernet module already occupies slot 0. The formula to calculate the interface number is:

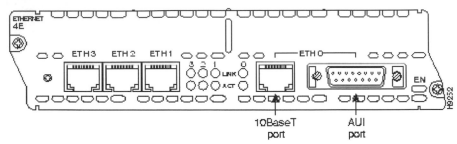

Access # config terminal
Access(config)#
Access(config)# ip routing
Access(config)# interface ethernet 0/0   ## The Ethernet module has been inserted in slop 0 and we are configuring the first ethernet port 'e0' in this network module card
Access(config-if)# ip address 207.80.127.130 255.255.255.192
Access(config-if)# ^Z   ## Exit the configuration mode
Access# wri mem
Access# show running   ## Shows the new running configuration

Figure 7.6 Cisco 3640 four-port ethernet network module and configuration.

Interface number = (32 × slot number) + unit-number + 1.

For example, the interface number for port 9 is (32 × 1) + 9 + 1 = 42. The range of interface numbers for our example is 33 to 64. The 68-pin interface connectors connect to an octal cable, which provides a 68-pin connector on one end and an eight serial (EIA-232) port connection on the other end. The eight serial port interface can be of RJ-45 type or DB-25 pin connectors. We will connect a modem at the end of each serial port.

Configuration Commands This configuration allows us to set up the asynchronous serial ports on the Cisco 3640 for secure dial-in access. Let's take a closer look at each one of the configuration commands shown in Table 7.2.

1. After you have entered the terminal configuration mode, enter the interface configuration mode for port 9. Port 9 maps to the interface number (32 × 1) + 9 + 1 = 42.

2. We can assign a dynamic IP address to each dial-in connection. Since there are only a small number of dial-in users, we will configure static IP addresses for each dial-in port. Notice the variable length subnet mask.

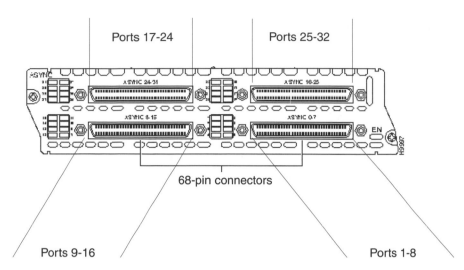

Ports 17-24   Ports 25-32

68-pin connectors

Ports 9-16   Ports 1-8

Figure 7.7 Cisco 3640 32-port asynchronous network module.

3. There are two options to setup the encapsulation mode, either SLIP or PPP. For security and flexibility in routing, it is best to set up the encapsulation to PPP.

4. Allows the PPP session to start automatically, when the access server receives the starting character.

Table 7.2
Cisco 3640 32-Port Asynchronous Network Module Configuration

1. *Access(config)# interface async 42*
2. *Access(config-if)# ip address 207.80.127.142 255.255.255.192*
3. *Access(config-if)# encapsulation ppp*
4. *Access(config-if)# autoselect ppp*
5. *Access(config-if)# ppp authentication chap*
6. *Access(config-if)# username jdoe password secretdoe*
7. *Access(config-if)# async mode dedicated*
8. *Access(config-if)# async default routing*
9. *Access(config-if)# line async 42*
10. *Access(config-line)# speed 64000*
11. *Access(config-if)# ^Z*   **## Exit the configuration mode**
12. *Access# wri mem*
13. *Access# show running*

5. Defines CHAP as the authentication protocol.

6. In our example, since the number of dial-in users is small, we will use the user-password configuration on the access server. Alternatively, we can configure a TACACS server and configure the authentication using TACACS by command "ppp use-tacacs" if desired.

7. The asynchronous dial-in line could be configured to be in dedicated or interactive mode. In the dedicated mode the user does not need to know about any configuration parameters like the IP address or the type of encapsulation, SLIP, or PPP. This is probably the best mode for supporting the dial-in access users.

8. Enables dynamic routing protocols on the asynchronous interface.

9. Enters the line configuration mode.

10. Configures the line speed.

11. Exits the configuration mode.

12. Writes the memory to the configuration.

13. Shows the running configuration.

This completes the configuration of the Cisco 3640 for dial-in remote access.

7.1.3   Conclusion

We started with the goal of creating an intranet implementation for NIT with a single-office configuration with secure dial-in access. We analyzed the requirements of NIT and designed a secure network architecture by isolating various functional departments and using appropriate ACLs based on the security policy. Then, we proceeded to design the secure dial-in access services with PPP/CHAP authentication on the asynchronous serial dial ports of the access server. NIT can provide for additional network security by deploying ACLs on the asynchronous interfaces as well.

In the next case study we will follow the expansion of NIT and its intranet to multiple branch offices.

7.2   Case Study II: Intranet in a Company Having Multiple Office Locations Spread Over a Wide Geographical Area

7.2.1   Case Study Objective

In this case study we will extend our study of a corporate intranet to include multiple geographical locations. The key aspects investigated here are the wide-area connectivity of multiple office locations in an intranet. As we have seen in Chapter 2, there are multiple technology options to implement a WAN, including the public frame relay network, the Internet, or the X.25 packet-switched network. This case study focuses on the use of a public frame relay network to implement NIT's expanded intranet. The configuration details for the network components are covered in depth.

7.2.2 Case Study Background and Requirements

NIT has just acquired a United States–based travel services company with offices in Boston; Washington, DC; and Chicago. The three offices are isolated islands of Windows NT–based ethernet LANs. Currently, the only means of communication between these offices is a nightly courier service. NIT would like to integrate these offices into its corporate intranet. In this case study we will implement NIT's expanded intranet using the frame relay technology, while in Case Study V we will implement the same intranet using the Internet-based VPN technology.

Frame Relay Network Architecture

We covered the frame relay technology in great detail in Chapter 2. Those of you who jumped directly to the case study section, please review the frame relay section before proceeding with this discussion. In this section, we will cover the requirements for NIT's frame relay network architecture and then design NIT's intranet network architecture based on these discussions.

Sizing Frame Relay Port Capacity and CIR Value

How do we size the frame relay port on the provider's network? There are two design choices, the first choice is based on an over-subscription model while the other choice is based on an under-subscription model. In Figure 7.8, the frame relay port connecting the HQ site needs to accommodate three PVCs with certain CIR values. If the PVCs connecting the regional offices to HQ have a CIR value of 64 Kbps, and if all were to burst to their CIR values at the same time, the port connecting HQ will need to have a port capacity of at least (64 Kbps $\times$ 3 = 192 Kbps). In this setup the design choices are:

- To get a port size of 192 Kbps;
- To get a port size greater than 192 Kbps;
- To get a port size smaller than 192 Kbps.

Why would anyone consider under-subscribing the port capacity? The answer lies in the fact that the statistical probability of all sites bursting to the

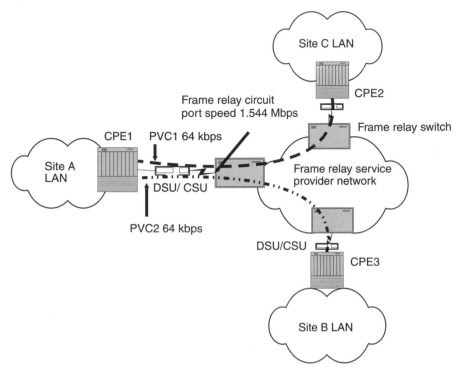

Figure 7.8 Frame relay network configuration parameters.

full CIR value is very low. Organizations can also take steps to ensure that the traffic-intensive activities between the regional sites and HQ are coordinated such that all daily batch transfers occur one after another and not at the same time. This certainly reduces the cost, but leaves less room for flexibility in the operation and future expansion. In most cases, the IT group has to take a guess at the traffic pattern and the capacity requirements. If you have a reasonable idea of the throughput requirements, the frame relay ports can be sized more accurately.

To illustrate how an educated guess at the capacity requirements could be made, let's consider some typical intranet applications in NIT's intranet. These applications include intranet Web access, Web-based ticketing transactions, Web-based search, and streaming of audio/video data in the future. The average size for a typical Web page with five graphics and text will be 60 KB. Let's say each regional office of NIT employs 50 users. If we assume that there are a total of 50 Web accesses in one minute, the capacity requirement at HQ would be 50 transactions per minute $\times$ (1/60 per sec) $\times$ 60 KB $\times$ 8 bits/byte = 400 Kbps. This example is an approximation and there would be an added overhead of TCP/IP, lost frames, and the resulting retransmissions. Based on the assumptions in this example, NIT would need to buy port capacity close to 400 Kbps at the HQ site or the next capacity increment available from service providers (i.e., 512 Kbps).

There is another technique to size a CIR value. This technique is based on the amount of time it takes to transmit a certain amount of data at a certain CIR value. Consider the example shown in Figure 7.9.

In this example, the Chicago site is connected to the frame network with a 64-Kbps circuit and has separate PVCs from Chicago to HQ and from Chicago to Boston. Each PVC has a 32-Kbps CIR. If Chicago were to transmit four transactions of 64 Kbytes each, and the circuit was able to burst to full circuit speed (i.e., at 64 Kbps), it would transmit these transactions in:

$$(4 \times 64 \text{ Kbytes} \times 8 \text{ bits/byte}) / (64 \text{ Kbps}) = 32 \text{ seconds}$$

If there is a new transaction between Chicago and HQ and the PVC from Chicago to Boston was being utilized to its capacity, the same four transactions will need to be transmitted on PVC to HQ at CIR value of 32 Kbps. In this case the transaction time would be $(4 \times 64 \text{ Kbytes} \times 8 \text{ bits/byte})/(32 \text{ Kbps}) = 64$ seconds. Is this an acceptable amount of delay for some of your mission-critical applications? If the answer is no, then you will have to resize the PVC to give the desired amount of transaction latency under any circumstances.

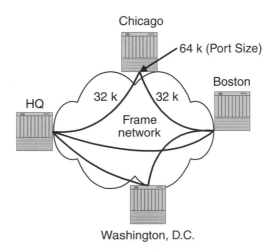

Partial mesh

Figure 7.9 Sizing CIR.

Network Topologies

How about the network topology? During our discussion on network topology and high resiliency in Chapter 2, we discussed the full mesh, partial mesh, and other options that provide high-resiliency network infrastructure. Figure 7.10 shows a logical representation of the three topology options.

Let's evaluate the design options against our design goals:

- To establish full connectivity between the regional offices and the HQ *at all times;*
- To establish full connectivity amongst the regional offices *as required.*

Which one of these arrangements makes the most sense? The answer is, of course, "It depends!" The final design choice depends on the needs of an organization, size of the network, and size of the budget. If there are small numbers of nodes and it is critical that all sites need to talk to all sites at all times, then the full mesh design will be the design of choice. Be aware that as the number of node increases, the number of required PVCs increases exponentially, thus increasing the cost and the management complexity. In most cases, one can reduce the cost and complexity by designing a partial mesh solution as shown in Figure 7.10(b). The beauty of partial mesh is that all sites can still reach each other over at least two separate paths, maintaining a primary and a secondary path of communication. The partial mesh solution will still result in a steep growth in the number of PVCs and the associated complexity, but a well-thought out placement of PVCs between critical sites can make this design manageable. The last design option is to build a hub-and-spoke arrangement shown in Figure 7.10(c). In this arrangement, though the regional offices are not directly connected, they can still communicate through the HQ, where the HQ router acts as the hub. Which arrangement meets the design goals of NIT?

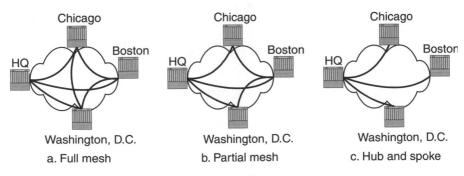

Figure 7.10 Frame relay network architecture options.

The first design goal is to establish full connectivity between the regional offices and the HQ *at all times*. The hub-and-spoke arrangement meets this goal of connectivity between the regional offices and the HQ. But this is not sufficient since we have to guarantee that the sites will be connected to the HQ at all times. What if the frame circuit connecting the Boston site to the frame cloud goes down? What if the hub router at the HQ site suddenly dies? In such scenarios we fail to meet the "at all times" requirement. We need to design a contingency solution(s) for these scenarios. Let's explore the second design goal.

The second design goal is to establish full connectivity among the regional offices *as required*. The hub-and-spoke arrangement certainly meets this requirement except in the case of failure in the hub router at the HQ. Based on this analysis, it is clear that "partial-mesh" arrangement meets the design goals of NIT. Next, we will consider the cost implications and various pricing options available from service providers.

Cost Analysis

In this section we will analyze the cost implications of design choices. In order to set up a frame relay–based intranet, NIT has to buy a frame relay circuit from its regional offices to a public frame relay network. In general, there are two types of pricing plans offered by frame relay service providers in the United States.

- Usage-based pricing;
- Fixed-rate pricing.

The general rule of thumb is that if you have a stable network configuration and you know your traffic pattern, you are better off choosing usage-based pricing. Here are some pricing-related tips to help you estimate the cost and do a price comparison.

- Check if the pricing is usage based or fixed rate. The usage-based pricing generally consists of charges per megabyte delivered and has a cap that corresponds to a fixed CIR. Any traffic above the fixed CIR value will be marked as discard eligible (DE).
- Estimate your monthly recurring cost. Your boss will probably like to have a general idea of the monthly cost involved.
- Ask for any hidden charges, like the cost for a local loop during the installation, any monthly surcharge, cost associated with PVCs such as a PVC reconfiguration charge.
- Ask if the PVC pricing is for a bidirectional or unidirectional CIR value.

- What are the charges if your traffic exceeds the CIR value and becomes DE?
- What are the maintenance charges? Is there a monthly or yearly maintenance plan?
- Is there any discount for certain number of circuits?

Sprint offers a zero CIR pricing. Although the pricing is attractive, this option should be carefully considered. If you know that the link is going to have a low utilization and is going to be used for nonmission-critical applications, this solution is quite attractive. Another situation where a zero CIR link will be useful is when the link is being considered as a backup link. There are creative solutions from other providers like MCI WorldCom that compete with Sprint. Here is an example.

MCI WorldCom offers usage-based pricing and charges 5 1/2 cents per megabyte of traffic sent within the CIR value. If the traffic exceeds the CIR value the frames are marked discard eligible (DE) and that portion of the traffic is charged at 4 1/2 cents. MCI WorldCom also charges a minimum usage fee for each PVC. In a recent price change, MCI WorldCom now charges only a $5 minimum usage charge per PVC for PVC of any CIR value up to 10 Mbps. This is attractive for NIT because NIT now can set up additional backup PVCs and pay only a monthly charge of $5 per PVC. This low pricing per PVC may be a good incentive for NIT to set up a partial mesh architecture instead of a hub-and-spoke architecture. As you can imagine, the pricing plans from service providers are in constant flux due to the competitive pressures. The prices quoted here may not be valid by the time this book goes to press, but the thought process will still be useful in conducting your cost analysis.

Resiliency

In order to provide a high-resiliency network infrastructure it is a good design exercise to consider failure scenarios for the link, hub, and router. These failure scenarios affect both the full mesh and partial mesh designs. First let's consider the scenario where a frame circuit connecting the Boston site to the frame cloud is down. There are two ways to deal with this scenario.

- To provision a backup secondary frame circuit with a lower CIR value between the Boston office and HQ;
- To set up a backup PSTN or ISDN dial-backup configuration.

The secondary frame circuit is a more expensive option. One can always be creative in using this secondary circuit as a traffic load balancing or overload protection circuit. In this example, the ISDN dial-backup configuration option offers the most cost-effective option without compromising the bandwidth requirements.

How about a failure in the hub router at the HQ site? The best way to deal with this issue is to add another router at HQ and terminate the backup frame circuit or the backup dial-back interface on this router. Internal LAN at HQ can also benefit from such a router. The LAN side of the primary and the secondary router will have to implement a configuration that provides intelligent routing to direct traffic from the failed router to the backup router. If the routers are Cisco routers, one way to implement such a scheme would be to set up an HSRP configuration on both routers.

Implementation Details

In this section we complete the discussion on the network architecture, the design parameters, and take a look at the device configurations. Let's start by looking at following design parameters:

- The network topology, frame relay circuit specifications, the number of PVCs, and their CIR values;
- The frame relay service provider and the port capacity;
- The frame router.

1. *Network topology:* Please refer to Figure 7.12 later in this section for the network architecture. Based on the cost and resiliency analysis, partial mesh network topology has been selected to provide the most cost-effective, high-resiliency option for NIT. The primary mesh consists of three 64-Kbps PVCs connecting the three regional offices to HQ in a hub-and-spoke arrangement. There are two backup 64-Kbps PVCs connecting the Chicago and Washington, DC, offices to the Boston office in a backup hub-and-spoke arrangement. Additional backup has been provided with an ISDN BRI (128 Kbps) dial-backup connection between Boston and HQ. This simple partial mesh arrangement provides connectivity between all offices in case of a failure in the frame circuit connecting HQ to the frame relay network.

2. *Service provider:* Based on the cost analysis, the SLAs, and other selection criteria, NIT has chosen MCI WorldCom as the frame relay service provider. NIT has also opted to use the usage-based pricing and reserve backup PVCs at a minimal usage fee of $5. HQ will be connected to the frame relay network with port capacity of 512 Kbps and each regional offices will be connected with a port capacity of 128

Kbps. Though the regional offices have a CIR of 64 Kbps, they may burst up to the full port speed of 128 Kbps. Thus, even in the case of a simultaneous burst from all regional offices, HQ can handle the traffic at a port capacity of 512 Kbps.

MCI WorldCom has also agreed to the following service guarantees:

- Circuit installations at all locations within 25 business days;
- Any request for port changes or changes to a PVC will be executed within one business day;
- In case of an outage, the mean-time-to-repair will be four hours;
- The frame relay circuits will exhibit a maximum delay of 70 milliseconds;
- In terms of frame delivery, MCI WorldCom will guarantee to pass 99.5% of the CIR traffic and 99% of the DE traffic;
- The network uptime is guaranteed to be 99.5%, which translates to a downtime of 3.6 hours per month.

If these SLAs are not met, MCI WorldCom will reimburse certain monthly charges to NIT. NIT has chosen to buy the CSU/DSU from MCI WorldCom as well. This is advantageous to both parties, since MCI WorldCom can now perform the periodic end-to-end test and management of the frame circuits from its Network Operations Center (NOC), without involving NIT resources. NIT will supply the frame router.

(*Disclaimer:* The price estimates and the SLA conditions were obtained from MCI WorldCom's Web site and press briefings and were not quoted by MCI WorldCom. These prices are indicative prices for the purposes of this case study and may not be available from MCI WorldCom.)

3. *Frame router:* The requirements for router processing power are different at HQ than at the regional offices. The HQ router needs to connect two internal 10BaseT ethernet LANs to the frame relay network and also act as a hub in the partial mesh configuration. This translates to a need for more processing power to process more traffic than the regional routers. In the previous case study we had selected a Cisco router and for the purposes of this case study we will continue to use the Cisco 4700 router with three network module (see Figure 7.11) as it meets all our design requirements for a router. The Cisco

4700 contains a 133-MHz Orion RISC microprocessor from Integrated Device Technology, Inc. (IDT) and is configured with:

- 32 MB of main memory. The main memory holds the router's running configuration and the all the routes in the routing table.
- 512 KB of secondary cache memory.
- 8 MB shared memory (DRAM). Shared memory is used for buffering packets arriving on various interfaces.
- 16 MB of flash memory. Flash memory holds the operating system (OS). It is good to get additional memory since there will be times when you would like to keep multiple copies of the OS in the memory. One example would be to perform router IOS (Inter-OS) upgrade. During an upgrade it is always good to have the original copy of the OS handy, in case you want to fall-back to the original version.
- 128 KB of nonvolatile RAM memory.
- 128 KB of boot ROM. The boot ROM contains a subset of Cisco IOS and is used to provide the bootstrapping of the router. When you first receive the router, the router will boot based on this Cisco boot program stored in this ROM.
- 8 MB boot flash.

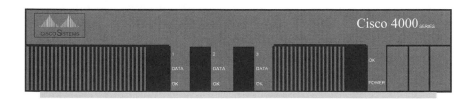

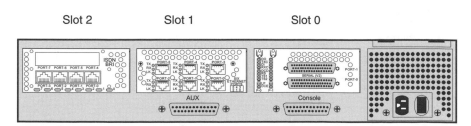

Figure 7.11 Frame relay router Cisco 4700.

Each Cisco 4700 router slot can accommodate a variety of network interface cards. The slot closest to the power supply is numbered slot 0. The slots going to the left of slot 0 are numbered as slots 1 and 2. The serial interface module has two EIA/TIA-449 serial ports that can accommodate connection speeds up to a T1. (The standard EIA-232 interface can support speeds up to 64 Kbps only.) Slot 1 has a six-port ethernet module with six RJ-45 interfaces, while slot 2 has a four-port ISDN BRI network interface card. The ISDN interface needs an external network terminating equipment called NT1 which is usually provisioned by the LEC providing the ISDN connection.

Consolidated Network Architecture

The architecture shown in Figure 7.12 consolidates all the design work in this case study and provides a WAN architecture for connecting all the regional offices to the HQ LAN in New York.

The architecture is a partial-mesh topology, connecting a Cisco 4700 router at all four sites using the frame relay network from MCI WorldCom. The Digital Link 3800 CSU/DSUs are provided by MCI WorldCom as well. The local loop is provided by MCI WorldCom wherever it is available, otherwise the LEC or CLEC provides the local loop from the NIT site to the nearest MCI WorldCom POP.

On the LAN side of the architecture, the ethernet interface e0 on CPE1 router connects to the administrative LAN, while the interface e1 is connected to the sales and customer service LAN. Figure 7.13 provides LAN architecture at NIT HQ.

Before we cover the configuration of network devices, it is important to cover the frame relay and IP addressing plan. Please refer to the architecture diagram shown in Figures 7.12, 7.13, and 7.14 for this discussion.

Frame Relay Addressing

When the service provider provisions PVCs, the provider also assigns DLCIs at each end of the PVC. The DLCIs act as local addresses on a frame relay PVC. The DLCIs have local significance only. The assignment of DLCIs is shown in Figure 7.14.

IP Addressing

There are various options for IP addressing. We will take a look at three IP addressing options. Each option deals with the issue of IP addressing associated with the PVCs connecting the four sites.

Option 1   In this option, we make use of the remaining address space from the 204.70.127.0 address space for providing addresses to the PVC end points.

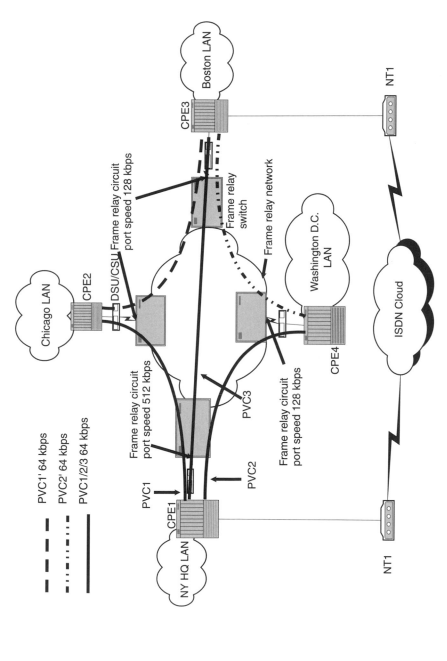

Figure 7.12 NIT WAN architecture for corporate intranet.

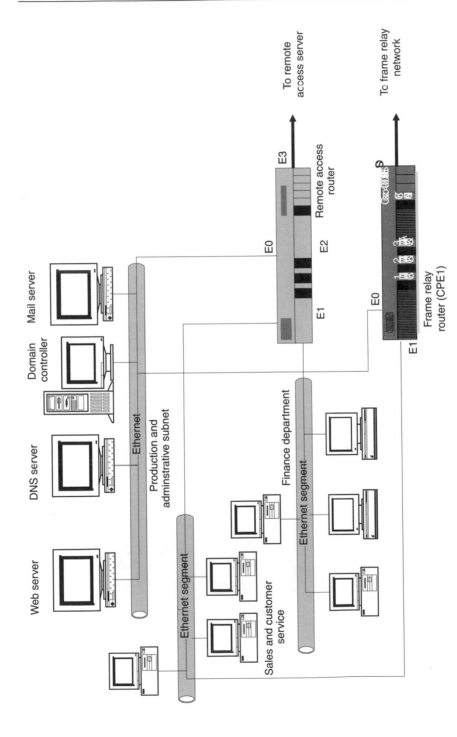

Figure 7.13 LAN network architecture at NIT HQ.

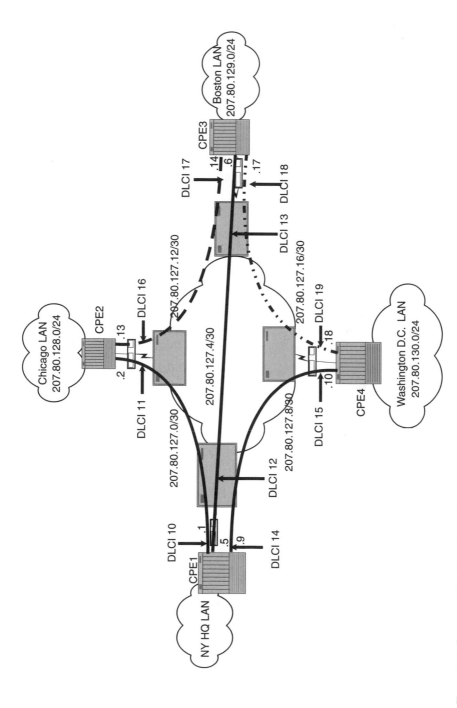

Figure 7.14 Frame and IP addressing scheme.

This class C address was used at the HQ in case study one. As for the regional office's LANs, we will assign a separate class C for each LAN as follows:

207.80.128.0/24	Chicago LAN
207.80.129.0/24	Boston LAN
207.80.130.0/24	Washington, DC LAN

In Case Study I, the 204.70.127.0/24 address space was variably subnetted into the subnets shown in Table 7.3.

The first 32 addresses have not yet been assigned. This block of 32 addresses is further subnetted into smaller subnets of only two host addresses by using a netmask of 255.255.255.252. The address allocation table is shown in Table 7.4.

The subnetted addresses can also be represented as 207.80.127.0/30, which translates to the .252 mask for the fourth octet in the address space. Figure 7.14 represents this addressing scheme and the DLCI addressing.

Option 2   Cisco offers another configuration option for creating IP-level connectivity between two sites over a point-to-point PVC. This Cisco option is called IP unnumbered. This configuration option allows us to save IP addresses allocated to the small IP subnet on each PVC as shown in Table 7.4. The IP unnumbered configuration is applied to each subinterface on the serial port S0 connecting to the frame network.

How does IP unnumbered work? Figure 7.15 shows NIT's intranet network topology with details on various serial and ethernet interfaces. With IP unnumbering, the serial port essentially borrows the IP address from the ethernet interface.

The partial configuration on the ethernet and the serial ports on CPE1 router are shown in Table 7.5.

Table 7.3
IP Addressing Scheme at HQ

Subnet	Netmask	Number of Hosts	Assigned To
207.80.127.32	255.255.255.224	32	Production
207.80.127.64	255.255.255.224	32	Sales and customer service dept.
207.80.127.96	255.255.255.224	32	Finance department
207.80.127.128	255.255.255.192	64	Remote access

Table 7.4
IP Subnets on the PVCs

Subnet	Netmask	Number of Hosts	Assigned To
207.80.127.0	255.255.255.252	2	HQ–Chicago PVC
207.80.127.4	255.255.255.252	2	HQ–Boston PVC
207.80.127.8	255.255.255.252	2	HQ–Washington, DC PVC
207.80.127.12	255.255.255.252	2	Chicago–Boston PVC
207.80.127.16	255.255.255.252	2	Boston–Washington, DC PVC

Option 3 If NIT does not need a full class C at each regional office and the number of hosts at each location is 50, then NIT can take the address space of 207.80.128.0/24 and variably subnet it into subnets of 64-host addressable addresses as shown in Table 7.6.

You can still use IP unnumbered configuration in this scenario, but it can only be used on the Chicago–Boston and Boston–Washington, DC link.

IP Routing Protocol

It's now time to select the routing protocol that will be run on all four routers to exchange the routing information between all four sites. The simplest case of routing between two sites is to use static routes. But static routes will not provide the dynamic reconfiguration that we would like to see in case of a link/PVC failure. This rules out use of static routes. If we consider use of routing protocols, the key requirement on such a protocol would be that it works in a variably subnetted environment. This rules out interior gateway protocols (IGP) like RIP and IGRP. The other IGP protocols that support variable length subnet masks (VLSM) are EIGRP, OSPF, RIP2, and IS-IS. EIGRP is a Cisco proprietary protocol and is limited to environment with Cisco routers. Since in this case study we are using all Cisco routers, we will configure the routers to run EIGRP protocol.

Configuration Details

We are now ready to cover the configuration details of the router and the remote access server. Please refer to the architecture diagram shown in Figure 7.14 for the rest of this discussion.

Configuration of the Cisco Routers Based on Option 1

Please refer to Table 7.7 for the configuration of the CPE1 frame relay router at the HQ in New York. Tables 7.8, 7.9, and 7.10 provide configuration for

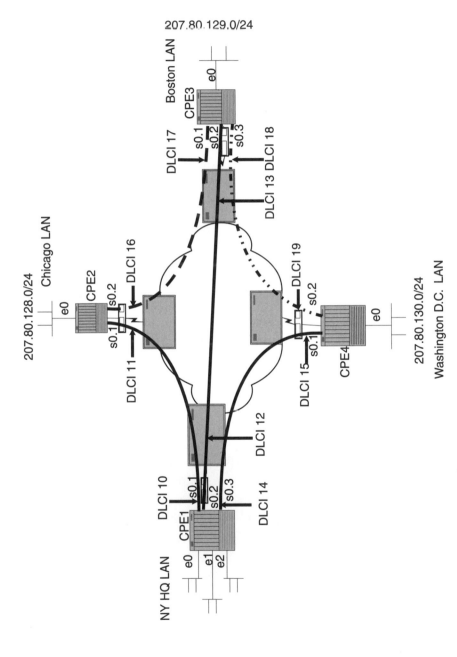

Figure 7.15 IP unnumbered addressing scheme.

Table 7.5
IP Unnumbered Configuration on CPE1

```
interface Serial0
description 512k frame link to MCIWorldCom
encapsulation frame-relay
interface s 0.1 point-to-point          ## The first sub-interface on serial port zero
ip unnumered ethernet 0
frame-relay interface-dlci 10 broadcast
!
interface s 0.2 point-to-point          ## The second sub-interface on serial port zero
ip unnumered ethernet 1
frame-relay interface-dlci 12 broadcast
!
interface s 0.3 point-to-point          ## The third sub-interface on serial port zero
ip unnumered ethernet 2
frame-relay interface-dlci 14 broadcast
!
interface Ethernet0                              ## This sets the ethernet interface
  ip address 207.80.127.33 255.255.255.224
                                  ## Serial interface s0.1 will borrow this IP address
  ip broadcast-address 207.80.127.63
!
interface Ethernet1                              ## This sets the ethernet interface
ip address 207.80.127.65 255.255.255.224
                                  ## Serial interface s0.2 will borrow this IP address
  ip broadcast-address 207.80.127.95
!
interface Ethernet2                              ## This sets the ethernet interface
  ip address 207.80.127.97 255.255.255.224
                                  ## Serial interface s0.3 will borrow this IP address
  ip broadcast-address 207.80.127.127
```

CPE2 (Chicago), CPE3 (Boston), and CPE4 (Washington, DC) routers. Please note carefully the configuration of serial and ethernet interfaces on each router.

Security Considerations

The security policy in this intranet configuration is based on the ACLs on the Cisco routers. For details on ACLs, please refer to Chapter 5 on security. The

Table 7.6
IP Subnets With One Class C Address Space

Subnet	Netmask	Number of Hosts	Assigned To
207.80.128.64	255.255.255.192	64	Chicago LAN
207.80.128.128	255.255.255.192	64	Boston LAN
207.80.128.192	255.255.255.192	64	Washington, DC LAN

policy adopted in this configuration is to set up most granular controls on the HQ ethernet interfaces on the CPE1 router. The ACLs 131, 132, and 133 on the interface e0, e1, and e2 allow only incoming connections from the address space associated with regional offices for services on secure (port 443) and unsecured (port 80) Web servers. Unrestricted access has been provided to the DNS (port 53), SMTP mail (port 25), and POP3 mail (port 110) services.

7.2.3   Conclusion

We expanded NIT's intranet from a single location to multiple locations. We have investigated the option of implementing secure intranet connectivity using a public frame relay service in detail. We also looked at the design parameters and the criteria used to make the final selection. We concluded with the detailed configuration of the routers and the routing protocol.

7.3   Case Study III: Intranet in a Company With a Legacy X.25 Connection to Its Operations in Europe

7.3.1   Case Study Objective

X.25 may be a legacy connection in the United States, but the X.25 protocol continues to be a dominant player worldwide. This case study illustrates various options available to network managers worldwide for interconnecting TCP/IP-based intranet using an X.25 network. Although we implement this interconnection using X.25 over frame relay, some of the methods discussed here can be used to deploy X.25 over on the Internet as well. In this case study, we finally set up NIT's Internet connection. In order to protect NIT's intranet from the public Internet, we will deploy a firewall to create the demilitarized zone. The configuration details of the intranet firewall server and the routers have been covered in great detail.

Table 7.7
Configuration of CPE1 Router (Cisco 4700) at NIT HQ

```
CPE1#wri t
Building configuration...
Current configuration:
!
! Last configuration change at 10:45:17 EDT Fri Jun 12 1998 by xxxx
! NVRAM config last updated at 17:20:20 EDT Mon Jun 15 1998 by xxxx
!
version 11.1
no service finger
no service pad
service timestamps debug datetime msec localtime show-timezone
service timestamps log datetime msec localtime show-timezone
service password-encryption
service compress-config
no service udp-small-servers
no service tcp-small-servers
!
hostname CPE1
!
clock timezone EDT-5
clock summer-time EDT recurring
boot system flash
boot system rom
enable secret 5 yyyy
enable password 7 ye4rftz3e2w
!
ip subnet-zero
no ip source-route
!
interface Serial0
description 512k frame link to MCIWorldCom
encapsulation frame-relay
interface s 0.1 point-to-point          ## The first sub-interface on serial port zero
ip address 207.80.127.1 255.255.255.252               ## To Chicago office
frame-relay interface-dlci 10 broadcast
```

Table 7.7 *(continued)*

```
!
interface s 0.2 point-to-point          ## The second sub-interface on serial port zero
ip address 207.80.127.5 255.255.255.252              ## To Boston office
frame-relay interface-dlci 12 broadcast
!
interface s 0.3 point-to-point           ## The third sub-interface on serial port zero
ip address 207.80.127.9 255.255.255.252          ## To Washington, DC office
frame-relay interface-dlci 14 broadcast
!
interface Serial1                             ## The interface is not being used
 no ip address
 shutdown
!
interface Ethernet0                         ## This sets the ethernet interface
 description Connection to the admin subnet
 ip address 207.80.127.33 255.255.255.224  ## Sets the interface IP address with the mask
 ip broadcast-address 207.80.127.63      ## The mask is a VLSM creating 32 host subnet
 ip access-group 131 out                    ## Sets the outbound ACL on this interface
 no ip redirects
 no ip proxy-arp
 no keepalive
!
interface Ethernet1
 description Sales and Customer service subnet
 ip address 207.80.127.65 255.255.255.224  ## Sets the interface IP address with the mask
 ip broadcast-address 207.80.127.95      ## The mask is a VLSM creating 32 host subnet
 ip access-group 132 out                        ## Sets the ACL on this interface
 no ip redirects
 no ip proxy-arp
 no keepalive
!
interface Ethernet2
 description Finance subnet
 ip address 207.80.127.97 255.255.255.224  ## Sets the interface IP address with the mask
 ip broadcast-address 207.80.127.127     ## The mask is a VLSM creating 32 host subnet
 ip access-group 133 out                        ## Sets the ACL on this interface
```

Table 7.7 *(continued)*

```
no ip redirects
no ip proxy-arp
no keepalive
!
!
interface Ethernet3
 no ip address
 shutdown
 !
   interface Ethernet4
     no ip address
   shutdown
!
interface Ethernet5
 no ip address
 shutdown
!
router eigrp 109                        ## Configures the EIGRP routing process
network 207.80.127.0                    ## Announces the route for the class C at HQ
!
access-list 131 permit tcp any any established
                                 ## Passes all the established connections
access-list 131 deny ip 207.80.127.32 0.0.0.31 207.80.127.32 0.0.0.31
                                            ## Anti-spoofing Filter
## This is also the first time you will observe the inverse masking technique used by
Cisco
access-list 131 permit tcp 207.80.128.0 0.0.0.255 any eq www
                    ## Allows all the Web connections on Port 80 from Chicago LAN
access-list 131 permit tcp 207.80.129.0 0.0.0.255 any eq www
                     ## Allows all the Web connections on Port 80 from Boston LAN
access-list 131 permit tcp 207.80.130.0 0.0.0.255 any eq www
                 ## Allows all the Web connections on Port 80 from Washington, DC LAN
access-list 131 permit txp 207.80.128.0 0.0.0.255 any eq 443
                ## Allows all the secure Web connections on Port 443 from Chicago LAN
access-list 131 permit tcp 207.80.129.0 0.0.0.255 any eq 443
                    ## Allows all the Web connections on Port 80 from Boston LAN
```

Table 7.7 *(continued)*

```
access-list 131 permit tcp 207.80.130.0 0.0.0.255 any eq 443
```
Allows all the Web connections on Port 80 from DC LAN
```
access-list 131 permit tcp any any eq 53
```
Allows all the DNS connections on TCP Port 53
```
access-list 131 permit udp any any eq 53
```
Allows all the DNS connections on UDP Port 53
```
access-list 131 permit tcp any any eq 25  ## Allows all the SMTP connections on Port 25
access-list 131 permit tcp any any eq 110
```
Allows all the POP3 connections on Port 110
```
access-list 131 permit icmp any any    ## Allows all the ICMP connections to the subnet
access-list 132 permit tcp any any established
```
Passes all the established connections
```
access-list 132 deny ip 204.80.127.64 0.0.0.31 204.80.127.64 0.0.0.31   ## Anti-spoofing Filter
access-list 132 permit tcp 204.80.129.0 0.0.0.255 any eq telnet
```
Allows all the telnet connections from Boston LAN
```
access-list 132 deny icmp any any
access-list 133 permit tcp any any established
```
Passes all the established connections
```
access-list 133 deny ip 204.80.127.96 0.0.0.31 204.80.127.96 0.0.0.31   ## Anti-spoofing Filter
access-list 133 deny icmp any any
!
  snmp-server community goodsnmp
!
line con 0
  exec-timeout 15 0
  password 7 yzcv45rfs6fgt$#eth%
!
line vty 0
  access-class 109 in                          ## Specifies who can login on the vty port
  exec-timeout 15 0
  password 7 6dtysb%3s8$frdtsr
!
  line vty 1 4
  access-class 109 in
  exec-timeout 15 0
  password 7 5thrs^srfv78fvsdt5t
```

Table 7.7 *(continued)*

!

end

7.3.2 Case Study Background and Requirements

With its success in the United States, NIT has decided to expand in Europe. As a result of this decision, NIT has acquired Prime Travel group, based in London. Prime's HQ in London is connected to its offices in Frankfurt, Amsterdam, Barcelona, and Paris using an X.25 network. Prime Travel will also provide NIT access to an IBM mainframe-based ticket reservation system for European airlines. To integrate Prime's operations into NIT's intranet, NIT has chosen to use a reliable frame relay connection to connect its IP-based intranet network to Prime's X.25-based network.

Further, to increase the company visibility and to boost ticket sales, NIT has decided to set up an Internet-accessible Web site in its New York office. NIT wants the Internet site to be well secured and isolated from the company's intranet.

Architecture

We can summarize NIT's requirements using a logical network architecture as shown in Figure 7.16.

As shown in Figure 7.16, the New York office acts as a gateway to the Internet as well as to the intranet connections in the United States and Europe.

The New York Office

In the previous two case studies we covered the design and architecture of the company intranet in the United States. The New York office acted as the hub for connectivity between other offices in the United States. Now to support the additional requirements of European expansion of the intranet and the Internet connection, the New York office needs to be re-engineered.

In order to support access to both the Internet and the intranet, we will design a network architecture that isolates the trusted internal network (the intranet) from the untrusted Internet. The network space separating the Internet and the intranet is popularly known as a DMZ (demilitarized zone). The DMZ is created using a firewall. The firewall device could be a server-based firewall (e.g., Checkpoint Firewall-1 software) or a router-based firewall. (Please refer to Chapter 5 on security for details on the types of firewalls and their vendors.)

For this design we will use Checkpoint's Firewall-1 product. The network setup with a firewall device is shown in Figure 7.17.

Table 7.8
Cisco Router Configuration for CPE2 Router

```
!
interface Serial0
description 128k frame link to MCIWorldCom
encapsulation frame-relay
interface s 0.1 point-to-point                          ## To HQ
ip address 207.80.127.2 255.255.255.252
frame-relay interface-dlci 11 broadcast
!
interface s 0.2 point-to-point                          ## To Boston
ip address 207.80.127.13 255.255.255.252
frame-relay interface-dlci 16 broadcast
!
interface Serial1                          ## The interface is not being used
 no ip address
 shutdown
!
interface Ethernet0                        ## This sets the ethernet interface
 description Connection to the local LAN
 ip address 207.80.128.0 255.255.255.0     ## Sets the interface IP address with the mask
 ip broadcast-address 207.80.128.255       ## The mask is a VLSM creating 32 host subnet
 ip access-group 131 out                   ## Sets the ACL on this interface
 no ip redirects
 no ip proxy-arp
 no keepalive
!
router eigrp 109                           ## Configures the EIGRP routing process
network 207.80.128.0                       ## Announces the route for the class C
!
access-list 131 permit tcp any any established
                                           ## Passes all the established connections
access-list 131 deny ip 207.80.128.0 0.0.0.255 207.80.128.0 0.0.0.255  ## Anti-spoofing Filter
access-list 131 permit ip any any          ## Allows all the connections
access-list 131 permit icmp any any        ## Allows all the ICMP connections to the subnet
```

Table 7.9
Configuration Router Details for CPE3 Router

```
!
interface Serial0
description 128k frame link to MCIWorldCom
encapsulation frame-relay
interface s 0.1 point-to-point                              ## To HQ
ip address 207.80.127.6 255.255.255.252
frame-relay interface-dlci 13 broadcast
!
interface s 0.2 point-to-point                           ## To Chicago
ip address 207.80.127.14 255.255.255.252
frame-relay interface-dlci 17 broadcast
!
interface s 0.3 point-to-point        ## The third sub-interface on serial port zero
ip address 207.80.127.17 255.255.255.252
frame-relay interface-dlci 18 broadcast
!
interface Serial1                        ## The interface is not being used
 no ip address
 shutdown
!
interface Ethernet0                       ## This sets the ethernet interface
 description Connection to the local LAN
 ip address 207.80.129.0 255.255.255.0    ## Sets the interface IP address with the mask
 ip broadcast-address 207.80.129.255      ## The mask is a VLSM creating 32 host subnet
 ip access-group 131 out                  ## Sets the ACL on this interface
 no ip redirects
 no ip proxy-arp
 no keepalive
!
router eigrp 109                          ## Configures the EIGRP routing process
network 207.80.129.0                      ## Announces the route for the class C
!
access-list 131 permit tcp any any established ## Passes all the established connections
access-list 131 deny ip 207.80.129.0 0.0.0.255 207.80.129.0 0.0.0.255   # Anti-spoofing Filter
access-list 131 permit ip any any         ## Allows all the connections
access-list 131 permit icmp any any    ## Allows all the ICMP connections to the subnet
```

Table 7.10
Configuration Details for CPE4 Router

```
!
interface Serial0
description 128k frame link to MCIWorldCom
encapsulation frame-relay
interface s 0.1 point-to-point                              ## To HQ
ip address 207.80.127.10 255.255.255.252
frame-relay interface-dlci 15 broadcast
!
interface s 0.2 point-to-point                            ## To Boston
ip address 207.80.127.18 255.255.255.252
frame-relay interface-dlci 19 broadcast
!
interface Serial1                           ## The interface is not being used
 no ip address
 shutdown
!
interface Ethernet0                         ## This sets the ethernet interface
 description Connection to the Local LAN
 ip address 207.80.130.0 255.255.255.0   ## Sets the interface IP address with the mask
 ip broadcast-address 207.80.130.255    ## The mask is a VLSM creating 32 host subnet
 ip access-group 131 out                        ## Sets the ACL on this interface
 no ip redirects
 no ip proxy-arp
 no keepalive
!
router eigrp 109                         ## Configures the EIGRP routing process
 network 207.80.130.0                      ## Announces the route for the class C
!
access-list 131 permit tcp any any established ## Passes all the established connections
access-list 131 deny ip 207.80.130.0 0.0.0.255 207.80.130.0 0.0.0.255  ## Anti-spoofing Filter
access-list 131 permit ip any any                  ## Allows all the connections
access-list 131 permit icmp any any    ## Allows all the ICMP connections to the subnet
```

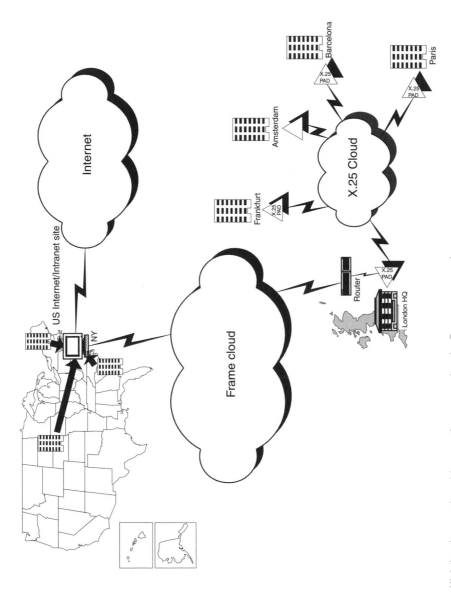

Figure 7.16 High-level network architecture for connecting the European operation.

The European Setup

The current setup in Europe is shown in Figure 7.18.

The server setup in the London office is shown in Figure 7.19.

The London office has two Sun Enterprise servers running the reservation server with access to the mainframe database. The Sun server–based reservation software has been written for the European operations and can accept information only in the X.25 format. The Sun servers are connected to the router using the SunLink X.25 Server Connect software. The servers use the on-board serial interface cards to set up the X.25 physical-level connectivity with the router. The on-board serial interface cards can support transmission speeds up to 64 Kbps. If one needs speeds higher than that, Sun offers a SunLink HSI (high-speed serial interface) card that can support speeds up to E1 (2.048 Mbps).

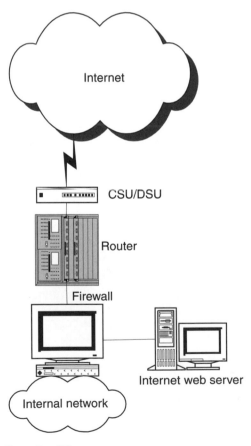

Figure 7.17 Internet firewall at HQ.

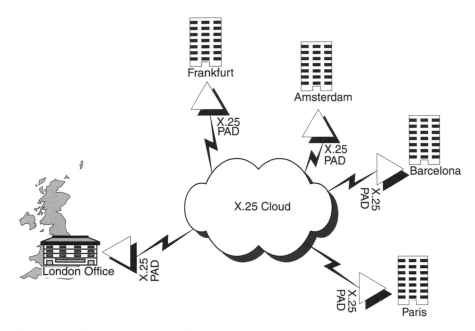

Figure 7.18 European network architecture.

New York to London Connectivity

The key challenge in expanding the United States–based intranet to Europe is to transport X.25 packets over a frame relay connection. We will explore the various approaches to accomplishing this task and provide a resilient architecture that provides the best response time.

Let's first explore some of the reasons behind the decision to use frame relay technology for this connectivity instead of a native X.25 connection between New York and London. Although from an implementation perspective a direct X.25 connection between New York and London would be an obvious choice, the technology has the following disadvantages:

- International X.25 connections are expensive.
- X.25 connections are not scalable enough to support future bandwidth-intensive applications.

While the frame relay connection offers following advantages:

- Low cost of ownership when using a public frame relay network;
- Ability to implement a high-resiliency network architecture at a reasonable cost;

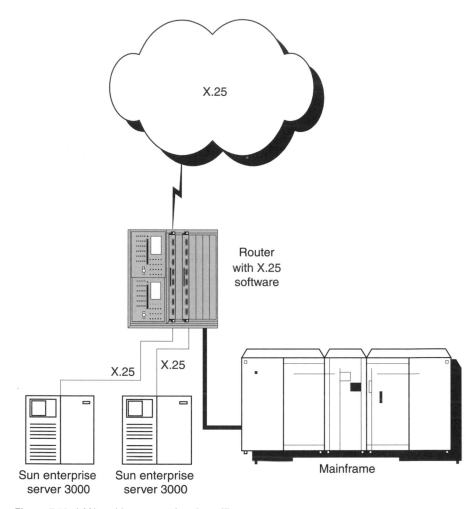

Figure 7.19 LAN architecture at London office.

- Ability to use the current investment in frame relay technology to support future TCP/IP-based applications;
- Scalable bandwidth.

The next challenge is to evaluate the various methods of transporting X.25 packets from New York to London over a frame relay circuit. The two most popular methods are:

1. Encapsulating X.25 packets in a frame and transporting the frame over the frame relay connection;

2. Encapsulating the X.25 packets in an IP packet and then transporting the IP packet over the frame relay connection.

The following sections cover both methods of transport in great detail.

Encapsulating X.25 Packets in a Frame

The best device to accomplish this task is known as FRAD (frame relay assembler/dissembler or frame relay access device). You can probably detect some similarity with the PAD (packet assembler/dissembler) concept from the X.25 world. FRAD performs functions similar to the PAD. FRAD provides the means to transport nonframe relay protocols like SNA, X.25, and IP over the frame relay network. Most of the FRAD manufacturers comply with RFC 1490, which sets the Internet standard for the multiprotocol interconnect over the frame relay. There are quite a few FRAD manufacturers in the market, including the market leaders Motorola and Cisco. The Motorola Vanguard series provides FRADs from the low- to high-end of the spectrum.

Selection of FRAD depends on following factors:

- Number of frame circuits supported by FRAD;
- Speed of the frame circuit supported;
- Whether the CSU/DSU is built into a FRAD or separate from it;
- The interface type for the frame termination: V.35, RS-232, or RS-449;
- Expansion capability of the FRAD;
- Types of LAN interfaces supported on the FRAD (e.g., Ethernet and/or token ring);
- Nonframe relay protocols supported by FRAD (e.g., IP, SNA, SDLC, BSC, or X.25).

Figure 7.20 shows the MP6520 FRAD from Motorola. The Motorola 6520 efficiently consolidates legacy SNA/SDLC BSC traffic with LAN traffic over dedicated or switched frame relay or an ISDN connection. The MP6520 router offers token ring and ethernet LAN support. MP6520 also includes support for two high-speed serial ports and three low-speed serial ports. The high-speed serial ports support speeds up to 1.544 Mbps. The two high-speed serial ports are ideal for frame relay connections. The MP6520 can be expanded to support a total of 19 serial ports. Additionally, MP6520 has the capacity to support multiple integral ISDN basic rate interfaces. The ISDN option can be used to setup the dial-backup connection, in case of a failure in the primary frame relay circuit.

Figure 7.20 MP6520 FRAD from Motorola.

Encapsulating X.25 Packets in an IP Packet

This method relies on encapsulating the X.25 packet in an IP packet and transmitting it over a frame relay network. Cisco provides the best example of implementing this methodology. The first part of transmitting X.25 over TCP (XOT) is documented in RFC 1613, and the transmission of IP packets over the frame relay circuit is documented in RFC 1490. RFC 1613 is based on Cisco's internal implementation of X.25 over a TCP engine also known as XOT. XOT allows for the transport of X.25 packets over an IP backbone infrastructure and is insensitive to the DTE/DCE role of the local interfaces at either end of an XOT TCP connection. XOT allows the router to tunnel X.25 VCs transparently over a TCP connection on a one-to-one mapping: one VC creates one TCP connection. Data and flow control packets are transported end-to-end. This implementation is symmetrical and requires that you have another RFC 1613–compliant device such as a Cisco router on the other side of the network to do the stripping of the TCP and IP header information before the delivery of X.25 packets to the remote X.25 device.

The transport of the IP packets over a frame relay connection occurs over the frame PVC that is set up by the frame relay service provider. The establishment of a X.25 PVC between two XOT entities occurs as a result of exchanging the previously mentioned nonstandard X.25 packets that are encapsulated in an XOT header. The X.25 PVC is established when the TCP connection has been completed. If the PVC fails to become established due to a network resource deficiency (for example, the destination interface is not up), then XOT will wait for a nonconfigurable internal timer to cycle through before it attempts to re-establish the connection. XOT PVC connections are never brought down unless the TCP connection is closed as a result of an abnormal event, such as bringing down the interface. For this case study we will use a Cisco router to maintain an all-IP, one-vendor environment.

Design of the Frame Circuit

Having decided on the end devices, let's take a look at the design of the WAN frame relay connection between the London and the New York offices. The primary requirement for the design of this circuit is to architect a highly responsive WAN connection that meets the desired resiliency requirements.

- The desire for a highly responsive frame connection needs to be quantified. The quantified design requirements will lead us to the specifications of the frame circuit. The specifications will define frame-circuit bandwidth parameters like the frame switch-port speed and the frame CIR.
- The resiliency requirements will lead us to a design that will incorporate backup strategies to account for the possibility of failure in any network component associated with the primary frame circuit. These network components could include the frame circuit itself, CSU/DSU, network terminating equipment (NTU), the frame switch in frame provider's network, or the CPE FRAD.

We will address the design and selection of frame parameters with requirements for a high-resiliency frame relay connection using either a backup frame circuit or a dial-back connection with ISDN or an asynchronous dial-up PSTN connection.

CIR and Frame Port

How does one quantify the notion of a "highly responsive" frame circuit? The first step in this process is to identify the traffic type and its characteristics. Please refer to Figure 7.21 for this discussion.

Initially, the traffic across the frame relay connection would consist of the X.25 request-response packets and IP request-response packets for intranet Web access. In either case, there are small number of request packets and a large number of response packets. These observations lead us to the first specification for the frame circuit. We need a frame circuit with the same CIR value in each direction. Most service providers provide the same bidirectional CIR value by default; if not, this will need to be specified. If the traffic volume exceeds the CIR value, the frames being transmitted are marked DE (i.e., the frame switches in the network will give priority to the frames that are not marked with the DE bit and in case of congestion drop the DE marked frames). It is the responsibility of the higher level protocols to take the appropriate actions to recover the lost data. A lost frame results in actions like retransmission by the higher level protocols. In the case of X.25, this is accompanied by a reduction

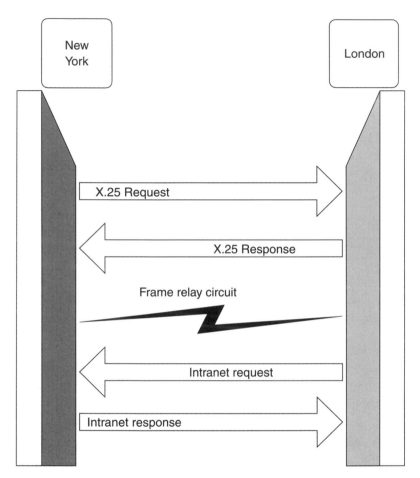

Figure 7.21 Traffic type over the New York–London frame connection.

in the receiving window size, using a sliding window mechanism. In principle, this mechanism is similar to the TCP flow-control mechanism. The net result is a reduction in the network throughput and thus the response time. To avoid this cascading effect, it is important that we choose a right value for the CIR.

NIT expects to receive a traffic burst of 64 Kbps in either direction; hence, to design a highly responsive frame connection, NIT ordered a PVC with a 128-Kbps CIR value in both directions. Since this is twice the expected traffic volume at any instance, this guarantees the packet transfer without a drop. This is also known as "over engineering" of the network.

The size of a frame port on a router depends on the number of PVCs that are supported by this port and their CIR values. If we are to support two PVCs, one for the current X.25 traffic and one for the IP traffic, we need to add the two CIR values and allow for future expansion. NIT expects the future IP traffic to have burst size up to 100 Kbps. Hence, looking at the current and the future needs, NIT decided to order a frame port of 256 Kbps.

Resiliency Design

The most resilient design will allow NIT to offer continuous services without traffic disruptions to its customers and employees. As mentioned earlier, the resilient design will need to account for the possibility of failure in any network component associated with the primary frame circuit. These network components could include the frame circuit itself, CSU/DSU, NTU, frame switch in frame provider's network, or the CPE FRAD. In this section we take a look at the backup strategy to meet the resiliency requirements using:

- A backup frame circuit;
- A backup ISDN connection;
- An asynchronous dial-up PSTN connection.

Figure 7.22 depicts all three backup strategies.

Backup Frame Circuit

One of the most effective ways to guard against the failure of a primary frame circuit is to set up a backup frame circuit. The backup frame circuit can provide an instantaneous fail-over in case of a failure in the primary circuit. The secondary backup circuit could be used in two ways: one as a standby circuit

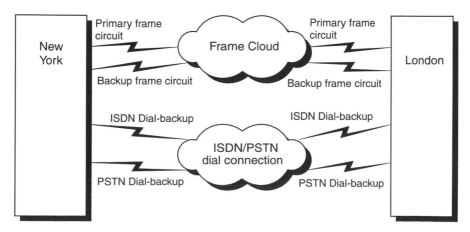

Figure 7.22 Backup strategies for high-resiliency network architecture.

and another as a load-balancing circuit. Figure 7.23 shows the two ways to set up a secondary backup circuit.

For router 1, both the primary and the secondary circuits terminate on the same router. In this case, it is quite easy to set up the secondary circuit in a standby or load-balancing mode. In this example we assume that the primary frame circuit was terminated on the serial interface s0, while the secondary was terminated on the serial interface s1. The following configuration on the s0 interface will achieve the fail-over and load sharing using the secondary circuit number 1 shown in Figure 7.23.

The configuration commands shown in Table 7.11 are explained below.

1. Enter the configuration mode.
2. Enter the interface configuration mode for serial interface s0.
3. Configure serial interface s1 as the backup interface for s0.
4. Configure the amount of time the secondary circuit should wait before taking over from the primary circuit and then giving the circuit back to the primary when the primary circuit is back on-line. (In this configuration we have configured the delay for the fail-over from primary (s0) to the secondary (s1) to be 60 seconds, while the fail-back

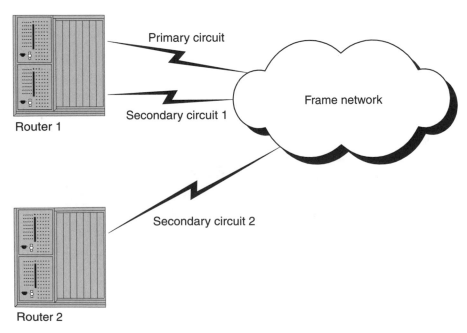

Figure 7.23 Backup using secondary frame circuit.

Table 7.11
Backup Frame Circuit Configuration

```
RouterR1# conf terminal
1.  RouterR1(config)# interface s0
2.  RouterR1(config-if)# backup interface serial 1
3.  RouterR1(config-if)# backup delay 60 never
4.  RouterR1(config-if)# backup load 70 30
5.  RouterR1(config-if)# ^z
6.  RouterR1# write mem
7.  RouterR1#
```

is configured to be never. The "never" value of the delay parameter for fail-back allows us to investigate the failure on the primary circuit and then bring the primary back to operation in a controlled manner.)

5. This configuration allows us to bring the secondary circuit on-line even if the primary is still up and running. With this command, a 70% usage on the primary circuit will set the trigger to bring the secondary circuit on-line. When the primary circuit utilization has fallen below the 30% mark, the secondary circuit will become idle again.

6. Exit the configuration mode.

7. Write the new configuration to memory.

8. Back to the enable prompt.

This setup will provide protection against the failure in a frame circuit, CSU/DSU, or NTU. If the primary and the secondary frame circuits are terminated on separate frame switches on the service provider's end, then you are protected against the switch failure as well. However, this solution does not protect you against a failure in router R1. The only protection against such failure would be to connect the secondary frame circuit to a separate router R2, as shown in Figure 7.23. Since most of the hosts in NIT's network would have the default router IP address of router R1, how do they sense a failure in router R1 and then switch their default gateway to router R2? This is a nontrivial question, but fortunately has vendor-proprietary answers. If you are using Cisco routers, you can implement a Cisco HSRP configuration to achieve this seamless transition. Please refer to the HSRP discussion in Chapter 3.

A natural question one might ask is, "What should the specifications of the backup frame circuit be?" If the backup frame circuit is going to be used in standby mode, then it is recommended to go with one-fourth the CIR value of the primary circuit. In some cases you might even get away with a zero CIR value, although you are obviously taking considerable risk with a zero CIR value, since you cannot guarantee any bandwidth to even the most critical business application(s). This may or may not be acceptable in an organization.

Frame Backup Using a Dial-Up ISDN or PSTN Connection

One of the most cost-effective ways of providing a backup connection is to use the dial-backup option. If the primary frame circuit goes down, the DTR signal on the backup serial interface goes high. This, in effect, signals the modem or the ISDN terminal adapter connected on the backup serial interface to make a dial-up connection to the ISDN or PSTN network.

The configuration of the router for a dial-backup connection is shown in Table 7.12.

The configuration commands shown in Table 7.12 are explained below.

1. Enter the terminal configuration mode.
2. Enter the configuration mode for the serial interface 0 (s0).
3. Assign IP address to the serial interface 0.
4. Specify the frame encapsulation type.
5. Configure to bring the backup interface up if the router detects the primary interface is down for more than 10 seconds. The router will maintain the backup connection until the primary connection is detected to be stable for 45 seconds.

Table 7.12
Dial-Backup Configuration

RouterR1# conf terminal
1. RouterR1(config)# interface serial 0
2. RouterR1(config-if)# ip address 132.44.32.35 255.255.255.0
3. RouterR1(config-if)# encapsulation frame relay
4. RouterR1(config-if)# backup delay 10 45
5. RouterR1(config-if)# backup interface serial 1
6. RouterR1(config-if)# interface serial 1
7. RouterR1(config-if)# ip unnumbered serial 0
8. RouterR1(config-if)# ^z
9. RouterR1# write mem
10. RouterR1#

6. Configure serial interface 1 (s1) to be the backup interface for s0.

7. Enter the configuration mode for the serial interface 1 (s1).

8. Configure s1 to pick up the IP address of s0 when s1 goes live (in this example, the backup interface goes live only if the primary interface s0 has failed).

9. Exit the configuration mode.

10. Save the new configuration to memory.

Note In the case of network equipment, and especially routers, the component most likely to fail is the power-supply module. The power-supply modules are highly susceptible to power surges; hence, it is recommended that the data center be protected against a power surge using uninterrupted power supply (UPS). Also, it is a good idea to order network equipment with dual power-supply modules.

Putting It All Together

The overall architecture with the Internet and intranet services is shown in Figure 7.24.

Figure 7.24 shows the various components discussed in this case study in great detail. We will first describe the overall architecture and then describe the specifications of each component. The New York site consists of the Internet and intranet Web clusters using the Sun Solaris platform running Netscape Web server software. The security at the site consists of the Cisco packet filters, Checkpoint Firewall-1 server, and secure OS configurations. The Internet site will have on average four to five graphics per page, but the total size of a Web page will be kept between 30–50 KB. The intranet site will have a similar type of content. The Internet site is expected to get 100,000 to 500,000 hits per day. The intranet site is expected to get 30,000 to 60,000 hits per day.

Component Details

All the major components in the architecture have been identified by an identification letter (e.g., R1 for router at New York). The following table identifies each code in Figure 7.24. The codes will be used to identify components in the rest of the design description.

We will now describe each component identified in the Table 7.13.

Component A1

Component A1 is a Sun cluster configuration with two Ultra Sparc servers. One can build a cluster of Sun servers using a Sun HA (high availability) solution, or alternatively, using a non-Sun solution like the Qualex HA solution

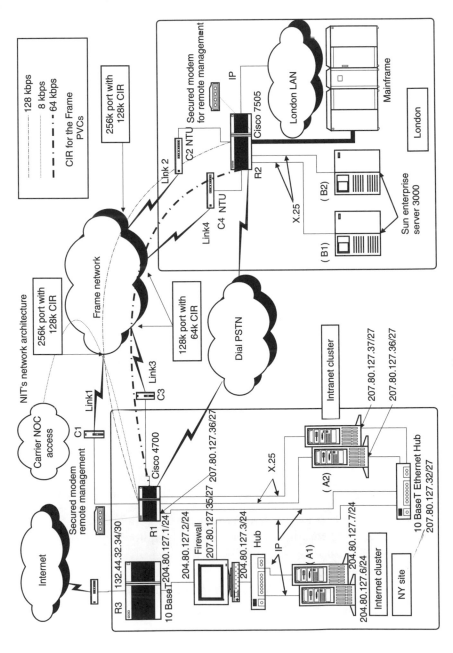

Figure 7.24 NIT's Internet and intranet network and security architecture.

Table 7.13
Intranet Architecture Components

Component Code	Description
A1	Sun Internet cluster
A2	Sun intranet cluster
R1	FRAD at New York
R2	FRAD at London
R3	Cisco router at NY
C1	CSU/DSU at NY
C2	NTU at London
C3	CSU/DSU at NY
C4	NTU at London
Firewall	Internet firewall
Link 1	Fractional T1 link from NY to the frame network
Link 2	Fractional E1 link from the London office to the frame network
Link 3	Backup fractional T1 link from NY to the frame network
Link 4	Backup fractional E1 link from the London office to the frame network
B1	Sun Enterprise Server 3000
B2	Sun Enterprise Server 3000

(URL). In either case, you will have a configuration as shown in Figure 7.25. The configuration requires:

- An S-Bus Quad Ethernet (four-port ethernet) card for the Sun servers;
- Two-port fiber channel or SCSI interface on each server for connection to the disk array.

In a server cluster the two servers exchange "keep alive" heartbeats at all times over the private LAN connection. If one of the servers detects that the other server has failed to announce the keep alive message, after a certain time-out period the server takes over the shared RAID5 disk set and the services of-fered by the cluster, including the IP address. In most of the cluster configura-tions, one server is active while the other is in standby mode. The active server is called the primary server, while the standby server is called the secondary server. Figure 7.25 shows the network connection for the heartbeat exchange. The two of the four ethernet ports on a quad ethernet interface card are used to directly connect the two servers. The two connections provide redundancy to protect against failure in the cables. The ports are 10 or 100 Mbps auto-sensing ethernet ports. Most of the HA software needs script development that allows the services (like the Web server or the database server) to fail-over from one server to another. The disk sets can be RAID5 or RAID 0+1 (i.e., two disk stripes are mirrored to form one disk set).

Component A2

The intranet cluster has a similar cluster configuration as the component A1. The intranet cluster differs significantly in its network interfaces. The cluster has a 10BaseT-ethernet connection to the firewall and a serial X.25 connection to the FRAD R1. The on-board serial interface of a Sun Ultra Sparc server can support serial transmissions up to 64 Kbps. If you need an interface speed higher than 64 Kbps, Sun also provides a high-speed serial interface card that can support transmission speeds up to 2 Mbps. Sun support for the X.25 protocol is via the Sun Server Connect product line.

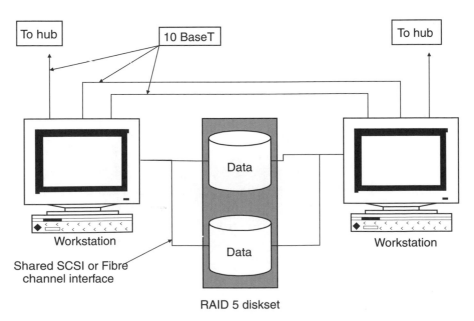

Figure 7.25 Sun cluster configuration for high availability.

Component R1

The routers R1 and R2 are critical components in this architecture. Configuration information for both Motorola and Cisco routers is given below:
Motorola 6520 MPRouter (FRAD) Standard Platform contains:

- 2 × high-speed serial interfaces (for the WAN side);
- 3 × standard serial interfaces (for the LAN side).
- 4 MB RAM;
- Option 1: ELAN (for the ethernet connectivity);
- Option 2: V.22bis modem (for dial-back backup);
- Option 3: 16 MB expansion;
- Option 4: 19-inch Rack Kit;
- Option 5: Redundant power supply (for added resiliency).

Cisco 4700 router comes with following options:

- 64 MB DRAM main memory;
- 16 MB secondary memory;
- 8 MB flash memory;
- 128 KB NVRAM;
- 8 MB boot flash;
- One NP-2E, 2-port ethernet;
- One NP-2T16S, 2-port synchronous serial high-speed serial interface and 16 synchronous/asynchronous low-speed serial interface (minimum IOS version 11.2(5)P required). (The high-speed ports supports EIA/TIA-232, EIA/TIA-449, V.35, X.21, NRZ/NRZI, DTE/DCE, or EIA-530 DTE interfaces. The low-speed ports support EIA/TIA-232, V.35, or X.21 interfaces in DTE or DCE mode. Each low-speed port can be individually configured for synchronous or asynchronous).

This leaves one slot for future expansion.
Another alternative router would be the Cisco 7505 or 7507 router. These routers have better expansion capabilities. The Cisco 7505 has a five-slot chassis, while the Cisco 7507 has a seven-slot chassis. The 75xx series routers also supports channel interface processor (CIP) from Cisco to connect directly to the mainframe. This is a highly cost-effective solution and avoids using a front-end processor (FEP). The Cisco 7500 series is much more powerful and is a

replacement series to the 7000 series. Hence, between the 7000 and 7500 series, it is best to choose the 7500 series router.

Considering the current and future needs of NIT, it is cost effective to invest in a Cisco 4700 router for R1 and a Cisco 7505 router for R2. This allows NIT to maintain one vendor infrastructure, thus reducing the cost of personnel, maintenance, and spares.

The routers R1 and R2 will have a secure modem connected to one of the serial interfaces. This allows the network management center of the Internet service provider to dial into the router for remote diagnostic and maintenance services. The two WAN interfaces of the routers support the physical connections to the frame circuits with the CSU/DSUs. These interfaces also support multiple logical frame connections, called the PVCs.

Component R2

The configuration for a Cisco 7505 is given below:

- 64 MB RAM;
- 16 MB flash SIMM;
- 8 MB boot flash;
- 128 Kbps NVRAM;
- The system will have four slots for interface processors and one slot for the system processor (RSP2) board;
- 6-port ethernet interface processor (EIP);
- 8-port fast serial interface processor (FSIP) with two port adapters. One port adapter contains four E1-G.703/G.704 ports supporting E1 (2.048 Mbps) circuits, and the second port adapter can support synchronous serial communications with EIA/TIA-232 interface.

Component R3

This component is a Cisco 4700 router with following configuration:

- 64 MB RAM;
- 8 MB RAM;
- Cisco IOS with IP feature set with support for BGP4;
- NP-2T, two serial;
- NP-6E, six-port ethernet interface.

The router is connected to the Internet with a dedicated serial connection of T1 bandwidth. The ACLs on this router allow for only port 80 connection to the Web server on the A1 cluster. All the rest of the ports have been denied access by the implicit denial in the Cisco ACL rule. The configuration of the Cisco router in Table 7.14 illustrates the setup.

Components of Link 1 and Link 2

Link 1 and link 2 will need to be provisioned with the appropriate carrier (e.g., MCI WorldCom in the United States with the MCI WorldCom Hyper Stream Frame Relay Network (HSFR), and British Telecom (BT) in the United Kingdom). The frame circuits need to be ordered with fractional T1 and E1 capacity at the United States and United Kingdom end respectively. As shown in Figure 7.24, the primary frame circuit will be ordered with a port capacity of 256 Kbps and a CIR of 128 Kbps. Link 1 will also support the network operation center (NOC) connectivity with an 8 Kbps PVC to router R1. The physical provisioning and installation of the frame circuits, especially international circuits, have a long lead time and should be accounted for in the project plan. The lead time could be sometimes as long as 30 to 45 business days.

Components of Link 3 and Link 4

Link 3 and link 4 act as backup links for links 1 and 2, respectively. The secondary circuit has a port capacity of 128 Kbps and a CIR of 64 Kbps in the United States and the United Kingdom. The FRADs at both ends need to be configured with floating static route configuration to allow for automatic rerouting of traffic.

Components of B1 and B2

The two Sun Enterprise servers act as the front-end gateways to the mainframe.

Configuration Details of Routers and Firewalls

For configuration of Router R3, see Table 7.14.

You can check on the ACL list applied to the ethernet interface by executing the following command at the login prompt, as shown in Table 7.15.

For configuration of Router R1, see Table 7.16.

For configuration of Router R2, see Table 7.17.

Firewalls

The firewall shown in Figure 7.24 protects NIT's Internet Web server cluster as well as the intranet LAN. Please refer to the firewall discussion in Chapter 5 for firewall-selection criteria. For the purposes of this case study we have selected Checkpoint's Firewall-1 software. Since this is our first use of the Checkpoint

Table 7.14
Router R3 Configuration

```
RouterR3#wri t
Building configuration...
Current configuration:
!
! Last configuration change at 10:45:17 EDT Wed Dec 24 1997 by xxxx
! NVRAM config last updated at 17:20:20 EDT Mon Dec 22 1997 by xxxx
!
version 11.1
no service finger
no service pad
service timestamps debug datetime msec localtime show-timezone
service timestamps log datetime msec localtime show-timezone
service password-encryption
service compress-config
no service udp-small-servers
no service tcp-small-servers
!
hostname routerR3
!
clock timezone EDT-5
clock summer-time EDT recurring
boot system flash
boot system rom
enable use-tacacs
enable last-resort password
enable secret 5 yyyy
enable password 7 ye4rftz3e2w
!
ip subnet-zero
no ip source-route
ip domain-list x.net
ip domain-list
clns routing
!
interface Serial0                    ## This defines the serial interface to the Internet
```

Table 7.14 *(continued)*

```
description T1 to AAST
ip address 132.44.32.34 255.255.255.252          ## This sets the interface IP address
ip broadcast-address 132.44.32.35
ip access-group 199 in                           ## This sets the ACL list on this interface
no ip redirects
no ip proxy-arp
ip route-cache sse
keepalive 40
!
interface Serial1                                ## The interface is not being used
no ip address
shutdown
!
interface Ethernet0                              ## This sets the ethernet interface
description Internet Web Server subnet
ip address 204.80.127.1 255.255.255.0     ## Sets the interface IP address with the mask
ip broadcast-address 204.80.127.255
ip access-group 132 out                          ## Sets the ACL on this interface
no ip redirects
no ip proxy-arp
no keepalive
!
interface Ethernet1
no ip address
shutdown
!
router bgp 4281                          ## Sets the BGP4 protocol on the serial Interface0
no synchronization
network 204.80.127.0                     ## Advertises the internal network with BGP
neighbor 132.44.32.33 remote-as 3561          ## Sets the ISP end of the serial link
neighbor 132.44.32.33 filter-list 1 out
!
ip name-server x.y.z.c
ip classless
ip as-path access-list 1 permit ^$
logging console informational
```

Table 7.14 *(continued)*

```
access-list 1 deny 0.0.0.0
access-list 1 permit any
access-list 109 permit ip host 204.80.127.2 any
access-list 132 permit tcp any any established
```
                                    ## Passes all the established connections
```
access-list 132 deny ip 204.80.127.0 0.0.0.255 204.80.127.0 0.0.0.255
```
                                            ## Anti-spoofing Filter
```
access-list 132 permit tcp any any eq www
```
                                    ## Allows all the web connections on Port 80
```
access-list 132 permit tcp any any eq 443
```
                                    ## Allows all the web connections on Port 443
```
access-list 132 permit icmp host 204.80.127.2 any
access-list 132 deny  icmp any any
access-list 199 deny  icmp any any
access-list 199 deny  ip 204.80.127.0 0.0.0.255 any
access-list 199 permit ip any any
!
  match as-path 6
  set local-preference 110
!
snmp-server community goodsnmp
!
line con 0
  exec-timeout 15 0
  password 7 yzcv45rfs6fgt$#eth%
line vty 0
  access-class 109 in
  access-class 109 out
  exec-timeout 15 0
  password 7 6dtysb%3s8$frdtsr
  line vty 1 4
  access-class 109 in
  access-class 109 out
  exec-timeout 15 0
  password 7 5thrs^srfv78fvsdt5t
  !
  end
```

Table 7.15
Checking Access Control Rules

```
RouterR3>sh access-list 132
  permit tcp any any established
  deny  ip 204.80.127.0 0.0.0.255 204.80.127.0 0.0.0.255
  permit tcp any any eq www
  permit tcp any any eq 443
  permit icmp host 204.80.127.1 any
  deny  icmp any any
```

Firewall-1 software, we will go through a detailed installation process and then create a firewall policy using the Firewall-1 GUI. The Firewall-1 software is available for various types of UNIX and Microsoft NT platforms. For our case study we will use the popular Sun Ultra Sparc2 workstation with dual CPU, 256 MB RAM, and 8 GB hard drive running Sun Solaris 2.5.1/2.6 Solaris OS. The size of RAM and the CPU processing power depends on the amount of line-rate traffic (e.g., 10/100 Mbps or OC-3 at 155 Mbps) that needs to be processed by the firewall. In general, the firewall processing introduces 2% to 3% overhead. If the firewall is responsible for VPNs with other Checkpoint firewall or SecuRemote dial-in clients, the processing power requirements will go up. In Figure 7.24, the firewall is connected to the internal LAN, the DMZ zone with the Internet Web server, and the Internet-connected router by 10BaseT ethernet interfaces. Sun Ultra Sparc comes standard with one 10- or 100-MB ethernet port on the system board. To accommodate an additional two ports, we can order the quad 10/100 PCI ethernet interface card.

The installation process below has been commented using the following fonts:

`(These are comments)`

To start the install process, run the "fwinstall" command from the compact disc containing the Firewall-1 software distribution. The command will have to be run by root or a userID with root privileges.

Note The installation process for Checkpoint Firewall-1 4.0 is similar to the one outlined beginning on page 293 for version 3.0b.

Table 7.16
Configuration Details for Router R1

R1#wri t
Building configuration...
Current configuration:
!
! Last configuration change at 10:45:17 EDT Fri Jun 12 1998 by xxxx
! NVRAM config last updated at 17:20:20 EDT Mon Jun 15 1998 by xxxx
!
version 11.2
no service finger
no service pad
service timestamps debug datetime msec localtime show-timezone
service timestamps log datetime msec localtime show-timezone
service password-encryption
service compress-config
no service udp-small-servers
no service tcp-small-servers
!
hostname R1
!
clock timezone EDT -5
clock summer-time EDT recurring
boot system flash
boot system rom
enable secret 5 yyyy
enable password 7 ye4rftz3e2w
!
ip subnet-zero
no ip source-route
!
x25 routing   **## Enable X.25 tunneling**
x25 route 10030024064 ip 166.42.44.6 166.42.44.10
    ## Creates an entry into X.25 routing table to forward any X.25 packets with the X.121 address to IP address specified destination. The two IP addresses are tried in sequence
service tcp-keepalives-in
service tcp-keepalives-out

Table 7.16 *(continued)*

```
!
interface Serial0
  description Primary Frame relay connection from NY to London
  no ip address
  encapsulation frame-relay
!
interface Serial0.1 point-to-point
  description Primary PVC over Link1 to London
  ip address 166.42.44.5 255.255.255.252
frame-relay interface-dlci 22
backup interface serial2
backup delay 40 never
!
interface Serial1
description Secondary Frame relay connection from NY to London
  no ip address
  encapsulation frame-relay
!
interface Serial1.1 point-to-point
  description Secondary PVC over Link2 to London
  ip address 166.42.44.9 255.255.255.252
  frame-relay interface-dlci 24
!
interface Serial1.2 point-to-point
  description Frame PVC to NOC
  ip address 166.42.44.13 255.255.255.252
  frame-relay interface-dlci 26
!
interface Serial2
  description dial-backup to London site
  ip unnumbered s0.1
  dialer in-band
  dialer wait-for-carrier-time 60
  dialer string 011441908660033
  pulse-time 1
  dialer-group 1
```

Table 7.16 *(continued)*

```
!
interface Serial3
 no ip address
 shutdown
!
interface Serial4
  description X25 connection to first Sun Servers at NY
  x25 pvc 1 tunnel 166.42.44.6 interface serial 4 pvc 2
                              ## To connect PVC's 1 and 2 across an IP network
!
 interface Serial5
description X25 connection to second Sun Servers at NY
 x25 pvc 3 tunnel 166.42.44.10 interface serial 5 pvc
                              ## To connect PVC's 3 and 4 across an IP network
!
interface Serial5.1 point-to-point
no ip address
 shutdown
!
interface Serial6
 no ip address
 shutdown
!
interface Serial7
 no ip address
 shutdown
!
interface Serial8
no ip address
shutdown
!
interface Serial9
no ip address
 shutdown
!
interface Serial10
```

Table 7.16 *(continued)*

```
no ip address
shutdown
!
interface Serial11
  no ip address
  shutdown
!
interface Serial12
  physical-layer async
  no ip address
  shutdown
!
interface Serial13
  no ip address
  shutdown
!
interface Serial14
  no ip address
  shutdown
!
interface Serial15
  no ip address
  shutdown
!
interface Serial16
  no ip address
  shutdown
!
interface Serial17
  no ip address
  shutdown
!
interface Ethernet0                              ## This sets the ethernet interface
  description Connection to the admin subnet
  ip address 207.80.127.36 255.255.255.224  ## Sets the interface IP address with the mask
  ip broadcast-address 207.80.127.63         ## The mask is a VLSM creating 32 host subnet
```

Table 7.16 *(continued)*

```
ip access-group 131 out                              ## Sets the ACL on this interface
no ip redirects
no ip proxy-arp
no keepalive
!
interface Ethernet1
description Sales and Customer service subnet
ip address 207.80.127.71 255.255.255.224  ## Sets the interface IP address with the mask
ip broadcast-address 207.80.127.95        ## The mask is a VLSM creating 32 host subnet
ip access-group 132 out                              ## Sets the ACL on this interface
no ip redirects
no ip proxy-arp
no keepalive
!
access-list 131 permit tcp any any established
                                    ## Passes all the established connections
access-list 131 deny ip 207.80.127.32 0.0.0.31 207.80.127.32 0.0.0.31  ## Anti-spoofing Filter
    ## This is also the first time you will observe the inverse masking technique used by
                                                                                 Cisco
access-list 131 permit tcp 207.80.128.0 0.0.0.255 any eq www
                        ## Allows all the Web connections on Port 80 from Chicago LAN
access-list 131 permit tcp 207.80.129.0 0.0.0.255 any eq www
                        ## Allows all the Web connections on Port 80 from Boston LAN
access-list 131 permit tcp 207.80.130.0 0.0.0.255 any eq www
                        ## Allows all the Web connections on Port 80 from DC LAN
access-list 131 permit tcp 207.80.128.0 0.0.0.255 any eq 443
                        ## Allows all the secure Web connections on Port 443 from Chicago LAN
access-list 131 permit tcp 207.80.129.0 0.0.0.255 any eq 443
                        ## Allows all the Web connections on Port 80 from Boston LAN
access-list 131 permit tcp 207.80.130.0 0.0.0.255 any eq 443
                        ## Allows all the Web connections on Port 80 from DC LAN
access-list 131 permit tcp any any eq 53
                        ## Allows all the DNS connections on TCP Port 53
access-list 131 permit udp any any eq 53
                        ## Allows all the DNS connections on UDP Port 53
access-list 131 permit tcp any any eq 25  ## Allows all the SMTP connections on Port 25
```

Table 7.16 *(continued)*

```
access-list 131 permit tcp any any eq 110
                              ## Allows all the POP3 connections on Port 110
access-list 131 permit icmp any any    ## Allows all the ICMP connections to the subnet
access-list 132 permit tcp any any established
                              ## Passes all the established connections
access-list 132 deny ip 204.80.127.64 0.0.0.31 204.80.127.64 0.0.0.31  ## Anti-spoofing Filter
access-list 132 permit tcp 204.80.129.0 0.0.0.255 any eq telnet
                              ## Allows all the telnet connections from Boston LAN
access-list 132 deny icmp any any
!
 snmp-server community goodsnmp
!
line con 0
 exec-timeout 15 0
 password 7 yzcv45rfs6fgt$#eth%
!
line vty 0
 access-class 109 in
 access-class 109 out
 exec-timeout 15 0
 password 7 6dtysb%3s8$frdtsr
!
 line vty 14
 exec-timeout 15 0
 password 7 5 thrs^srfv78fvsdt5t
 !
 end
```

```
# ./fwinstall

************** FireWall-1 v3.0 Installation **************

Reading fwinstall configuration.  This might take a while.
Please wait.

Configuration loaded.  Running FireWall-1 Setup.
```

Table 7.17
Configuration Details for Router R2

```
R2#wri t
Building configuration...
Current configuration:
!
! Last configuration change at 10:45:17 EDT Fri Jun 12 1998 by xxxx
! NVRAM config last updated at 17:20:20 EDT Mon Jun 15 1998 by xxxx
!
version 11.2
no service finger
no service pad
service timestamps debug datetime msec localtime show-timezone
service timestamps log datetime msec localtime show-timezone
service password-encryption
service compress-config
no service udp-small-servers
no service tcp-small-servers
!
hostname R2
!
clock timezone EDT-5
clock summer-time EDT recurring
boot system flash
boot system rom
enable secret 5 yyyy
enable password 7 ye4rftz3e2w
!
ip subnet-zero
no ip source-route
!
x25 routing
x25 route 20052464027 ip 166.42.44.5 166.42.44.9
service tcp-keepalives-in
service tcp-keepalives-out
!
interface Serial1/0
```

Table 7.17 *(continued)*

```
description Primary Frame relay connection from London to NY
no ip address
encapsulation frame-relay
!
interface Serial1/0.1 point-to-point
  description Primary PVC over Link1 to NY
  ip address 166.42.44.6 255.255.255.252
  frame-relay interface-dlci 23
  backup interface serial 1/6
  backup delay 40 never
```
Configures the backup line to take over after the primary has been down for 40 seconds. The "never" option prevents the backup circuit giving up to primary automatically when the primary circuit comes back on-line. This prevents a potential flapping situation and allows administrators to conduct full check on primary before bringing primary on-line and disabling the backup manually.

```
!
interface Serial1/1
description Secondary Frame relay connection from London to NY
  no ip address
  encapsulation frame-relay
!
interface Serial1/1.1 point-to-point
  description Secondary PVC over Link2 to London
  ip address 166.42.44.10 255.255.255.252
  frame-relay interface-dlci 25
!
interface Serial1/2
  no ip address
  shutdown
!
interface Serial1/3
  no ip address
  shutdown
!
interface Serial1/4
  description X25 connection to first Sun Servers at London
```

Table 7.17 *(continued)*

```
 x25 pvc 2 tunnel 166.42.44.5 interface serial 4 pvc 1
   ## Configures an X.25 tunnel connecting PVC's 2 and 1 over serial interface 4 with IP
   address of 166.42.44.5
!
 interface Serial1/5
description X25 connection to second Sun Servers at London
 x25 pvc 4 tunnel 166.42.44.9 interface serial 5 pvc 3
   ## Configures an X.25 tunnel connecting PVC's 4 and 3 over serial interface 4 with IP
   address of 166.42.44.9
!
interface Serial6
 no ip address
 shutdown
!
interface Serial7
 no ip address
 shutdown
!
interface Ethernet2/0                                    ## This sets the ethernet interface
 description Connection to the Local LAN
 ip address 208.80.127.1 255.255.255.0    ## Sets the interface IP address with the mask
 ip broadcast-address 208.80.127.255      ## The mask is a VLSM creating 32 host subnet
 ip access-group 131 out                          ## Sets the ACL on this interface
 no ip redirects
 no ip proxy-arp
 no keepalive
!
interface Ethernet2/1
no ip address
 shutdown
!
interface Ethernet2/2
no ip address
 shutdown
!
interface Ethernet2/3
```

Table 7.17 *(continued)*

```
no ip address
 shutdown
!
interface Ethernet2/4
no ip address
 shutdown
!
interface Ethernet2/5
no ip address
 shutdown
!
access-list 131 permit tcp any any established
```
                                    ## Passes all the established connections
```
access-list 131 deny ip 208.80.127.32 0.0.0.0 208.80.127.0 0.0.0.0     ## Anti-spoofing filter
```
   ## This is also the first time you will observe the inverse masking technique used by Cisco.
```
access-list 131 permit tcp any any eq www
```
                        ## Allows all the web connections on Port 80 from Chicago LAN
```
access-list 131 permit tcp any any eq 25   ## Allows all the SMTP connections on Port 25
access-list 131 permit tcp any any eq 110
```
                                ## Allows all the POP3 connections on Port 110
```
access-list 131 permit icmp any any     ## Allows all the ICMP connections to the subnet
access-list 132 permit any any established     ## Passes all the established connections
!
 snmp-server community goodsnmp
!
line con 0
 exec-timeout 15 0
 password 7 yzcv45rfs6fgt$#eth%
!
line vty 0
 access-class 109 in
 access-class 109 out
 exec-timeout 15 0
 password 7 6dtysb%3s8$frdtsr
!
```

Table 7.17 (continued)

```
line vty 1 4
  exec-timeout 15 0
  password 7 5thrs^srfv78fvsdt5t
!
end
```

Please read the following license agreement.
Hit 'ENTER' to continue...

        SOFTWARE RIGHT-TO-USE LICENSE AGREEMENT

This is the Software License Agreement between "You" and
Check Point Software Technologies Ltd. ("Check Point") of
3A Jabotinsky St., Ramat Gan, Israel.

. . . .
(Standard licensing terms and conditions go here.)
. . . .

Do you accept all the terms of this license agreement
(y/n) ? y
Checking available options. Please wait...................

Which of the following FireWall-1 options do you wish to
install/configure ?
----
(1) FireWall-1 Enterprise Product
(2) FireWall-1 Single Gateway Product
(3) FireWall-1 Enterprise Management Console Product
(4) FireWall-1 FireWall Module
(5) FireWall-1 Inspection Module

Enter your selection (1-5/a): 1

(The Enterprise Product contains all the available options
and has a license to protect unlimited number of enterprise
hosts. The single gateway product is available for

25/50/150 hosts. The management console, the firewall module and the inspection module can be loaded in isolation)

```
Installing/Configuring FireWall-1 Enterprise Product.

Which Component would you like to install ?
-----------------------
(1) FireWall & Management Modules
(2) FireWall Module only
(3) Management Module only

Enter your selection (1-3/a) (1):
Please wait...

Selecting where to install FireWall-1
--------------------
FireWall-1 requires approximately 10429 KB of free disk
space.
Additional space is recommended for logging information.

Enter destination directory (/etc/fw)): ↵

Checking disk space availability...

Installing FW under /etc/fw (345368 KB free)
Are you sure (y/n) (y) ? ↵

Software distribution extraction
-----------------
Creating directory /etc/fw
Extracting software distribution. Please wait ...
Software Distribution Extracted to /etc/fw
Installing license
---------------------------

Reading pre-installed license file fw.LICENSE... done.

The following evaluation License key is provided with this
FireWall-1 distribution
Eval            150ct97 3.x controlx pfmx routers connect
motif embedded
```

Do you want to use this evaluation FW-1 license (y/n) (y)?

*(For the initial installation use the standard bundled
software. We will install a valid license key after the
software installation has been completed)*

Using Evaluation License String

This is FireWall-1 Version 3.0b (VPN) (13Jul98 21:34:36)

Type   Expiration Ver Features
Eval   15Oct97   3.x controlx pfmx routers connect
motif embedded

License file updated
Module /etc/fw/modules/fwmod.5.3.o was updated

********* FireWall-1 kernel module installation *********

installing FW-1 kernel module...
Done.

Do you wish to start FireWall-1 automatically from
/etc/rc3.d (y/n) (y) ?

*(Choose Yes! This way you are protected in case of a sud-
den reboot of the system. As the system is going through
its boot process it will read the files in rc1.d, rc2.d and
rc3.d)*

FireWall-1 startup code installed in /etc/rc3.d

Welcome to FireWall-1 Configuration Program
===
This program will guide you through several steps where
you will define your FireWall-1 configuration. At any
later time, you can reconfigure these parameters by running fwconfig

Configuring Licenses...

```
========================
The following licenses are installed on this host:
Eval            15Oct97     3.x controlx pfmx routers con-
nect motif embedded

Do you want to add licenses (y/n) (n) ?
Module /etc/fw/modules/fwmod.5.5.1.o was updated

Configuring Administrators...
==============================
No FireWall-1 Administrators are currently defined for this
Management Station.

Do you want to add users (y/n) (y) ?
User: fwadmin
Permissions            ((M)onitor-only,(R)ead-only,(U)sers-
edit,read/(W)rite): w
Password:
Verify Password:
User fwadmin added successfully
```

(This userID will be used to connect to the firewall soft-ware to configure the firewall policy and monitor the fire-wall logs)

```
Add another one (y/n) (n) ?

Configuring GUI clients...
===========================
GUI clients are trusted hosts from which FireWall-1
Administrators are
allowed to log on to this Management Station using
Windows/X-Motif GUI.

Do you want to add GUI clients (y/n) (y) ?
Please enter the list hosts that will be GUI clients.
Enter hostname or IP address, one per line, terminating
with CTRL-D or your EOF character.
166.49.95.122
166.49.95.123
Is this correct (y/n) (y) ?
```

(The list of names or IP addresses configured in this step will be the only authorized clients allowed to use the "Policy Manager GUI")

Configuring Remote Modules...
==============================
Remote Modules are FireWall or Inspection Modules that are going to be controlled by this Management Station.

Do you want to add Remote Modules (y/n) (y) ? n

(Since in our example there is only one firewall, this option need not be configured)

Configuring Security Servers...
=================================

Installing Secured Services option

You may make the services below secured. By doing that you will enable the usage of strong authentication and/or content security for this service (see the user guide for details).

1) FTP
2) HTTP
3) TELNET
4) RLOGIN
5) SMTP

Please enter the numbers of the services you want to make secured.
For example, if you wish to make only ftp and rlogin secured, enter '1 4' > 4

You have selected rlogin to be secured.
Is this correct (y/n) (n): y

(These security server configurations create proxy servers on the firewall)

Do you wish to enable the client authentication feature
(y/n) (y) ? n

*(This authentication of client is based on RSA key ex-
change)*

In version 2.1, the FTP, Telnet and HTTP security daemons
were listening on their original TCP ports (i.e., 21, 23
and 80 respectively).
Beginning at version 3.0, these daemons may be installed
on randomly selected high ports instead. The FireWall gateway will redirect connections coming to the original ports
to these high ports.

Answering 'yes' to the following question will keep the
FTP, Telnet and HTTP security daemons on their original TCP
ports. This will allow security policies to be loaded from
version 2.1 control module on a version 3.0 gateway module.

Do you wish to enable backward compatibility, i.e. let 2.1
control stations control 3.0 inspection modules (y/n) (n)
?

Restarting the inet daemon (process 124)

Configuring Groups...
=====================
FireWall-1 access and execution permissions

Usually, FireWall-1 is given group permission for access
and execution. You may now name such a group or instruct
the installation procedure to give no group permissions to
FireWall-1. In the latter case, only the Super-User will
be able to access and execute FireWall-1.

Please specify group name (<RET> for no group permissions):
staff

Group staff will be used. Is this ok (y/n) (y) ?

Setting Group Permissions... Done.

Configuring IP Forwarding...
============================

Do you wish to disable IP-Forwarding on boot time (y/n) (y)
?
IP forwarding disabled

(IP forwarding should be strictly disabled. The firewall will then inspect each incoming or outgoing packet against the firewall policy and process it accordingly)

Configuring Default Filter...
=============================

Do you wish to modify your /etc/rcS.d boot scripts to allow a default filter to be automatically installed during boot (y/n) (y) ?

Which default filter do you wish to use?
----
(1) Allow only traffic necessary for boot
(2) Drop all traffic

Enter your selection (1-2) (1):

Generating default filter

Configuring Random Pool...
==========================
You are now asked to perform a short random keystroke session.
The random data collected in this session will be used for generating Certificate Authority. RSA keys.

Please enter random text containing at least six different characters. You will see the '*' symbol after keystrokes that are too fast or too similar to preceding keystrokes. These keystrokes will be ignored.

Please keep typing until you hear the beep and the bar is
full.

(   )   (.....................)

Thank you.

Configuring CA Keys...
=======================
fw: no license for 'ca'
The installation procedure is now creating an FWZ
Certificate Authority Key for this host. This can take several minutes. Please wait...
fw: no license for 'ca'

Configuration ended successfully

************** FireWall-1 is now installed. **************

Do you wish to start FW-1 now (y/n) (y) ?

Note: On first startup, Security Policy fetch error can be
IGNORED

FW-1: Starting fwd
fwd: Cannot establish Log Server on port 257: Address already in use
FW-1: Starting snmpd
snmpd: Opening port(s): 161 Cannot bind: Address already
in use
260 Cannot bind: Address already in use

SNMPD: server running
FW-1: Starting fwm (Remote Management Server)
fwm: Can't establish service: Address already in use
FW-1: failed to start fwm

FW-1: Fetching Security Policy from localhost
Trying to fetch Security Policy from localhost:
Failed to Load Security Policy: No State Saved

```
Fetching Security Policy from localhost failed
FW-1 started

*************************************************
                DO NOT FORGET TO:
1. add the line:    setenv FWDIR /etc/fw    to .cshrc
             or  FWDIR=/etc/fw; export FWDIR to .profile
2. add  /etc/fw/bin   to path
3. add  /etc/fw/man   to MANPATH environment
***********************************************************

You may configure FireWall-1 anytime, by running fwconfig.

********** Installation completed successfully **********

# fwstop

Uninstalling Security Policy from all.all@simba
Done.
```

(Now install the license string issued for your purchase of the Firewall software using 'fw putlic' command)

```
# fw putlic Ox<hostid> License string issued by Checkpoint
<Feature set ordered>

This is FireWall-1 Version 3.0b (VPN) (13Jul98 21:41:11)

Type                 Expiration Ver Features
ID-hostid            6Aug98    3.x controlx pfmx routers des
skip connect motif embedded
Eval                 15Oct97   3.x controlx pfmx routers con-
nect motif embedded

License file updated
Module /etc/fw/bin/../modules/fwmod.5.5.1.o was updated
```

(Start the firewall daemon)

```
# fwstart
FW-1: Starting fwd
```

```
fwd: FireWall-1 server is running
FW-1: Starting snmpd
snmpd: Opening port(s): 161
SNMPD: server running
FW-1:  Starting fwm (Remote Management Server)
fwm: FireWall-1 Management Server is running

FW-1: Fetching Security Policy from localhost
Trying to fetch Security Policy from localhost:
Failed to Load Security Policy: No State Saved

(This is OK, since we haven't created a security policy
yet!!)

Fetching Security Policy from localhost failed
FW-1 started
```

Now we are ready to create the firewall policy. The first step is to identify the goals of the firewall by answering the question "What are we trying to protect?" In the case of NIT's intranet, we are trying to protect NIT's internal LAN, the NIT intranet setup, and the DMZ with the Internet Web-server cluster. Our first step could be a very simple policy that allows ports 80 and 443 connections to the Web-server cluster and blocks access to everything else. For administration purposes, NIT may want to give its security personal access to the firewall management module over the Internet and log these access activities in detail. We will create a firewall policy reflecting these goals using the Security Policy management interface. The management interface is a GUI interface that is available for Windows 95/98/NT or UNIX workstations. We again use a popular platform, for this example, Windows 95.

Before we can create the policy we need to define various network objects that will be subjected to these policies. The network objects in Figure 7.24 that need to be defined are the individual servers in the Web-server cluster, the firewall server, and the router. The "network object" option is available under Manage → Network Objects menu. Please refer to Figure 7.26 for a sample screen image. Here, we are defining the firewall server and its internal interface IP address. The Interface tab defines the other two interfaces on the firewall. In case a company does not have enough IP addresses and is forced to use private (unregistered) IP addresses on their intranet, then the firewall could also act as a network address translator. Please refer to Chapter 5 for more details on NAT.

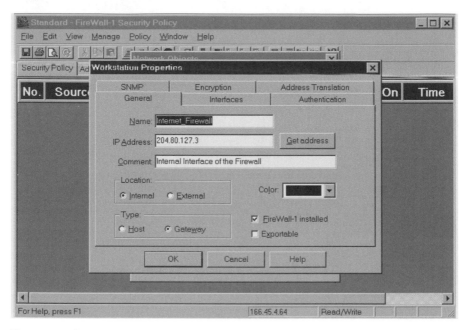

Figure 7.26 Creating network objects in Firewall-1.

Although these screen shots are in black and white, the actual screens are available in color. Figure 7.27 shows the various network objects that we need to create NIT's security policy.

Finally, we are ready to create and implement the firewall policy. Figure 7.28 shows the security policy that will be implemented for this firewall.

The rules in this policy follow our earlier discussion and meet the goals of setting up the firewall. Any packets that do not satisfy the first three rules are dropped and the information is logged in the log file. Web clusters 1 and 2 denote the Internet Web servers in the DMZ.

This completes our implementation of the NIT intranet for this case study.

7.3.3   Conclusion

This case study provides detailed configuration of new components such as the firewall and the X.25 connection to the legacy network. We took a close look at the various options available to transport the legacy traffic between NIT's North American and European operations. The next case study expands on the legacy connectivity requirements by setting up NIT's access to a legacy mainframe system in London by transporting SNA traffic over frame relay network.

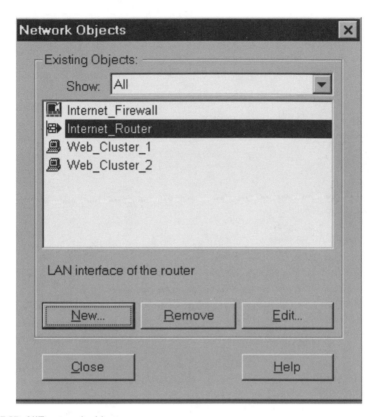

Figure 7.27 NIT network objects.

7.4 Case Study IV: Intranet in a Company With a Legacy Connection to an IBM Mainframe System Using SNA

7.4.1 Case Study Objective

At the end of this case study you will be able to set up an intranet connection to a legacy mainframe system using SNA over frame. The case study will provide the configuration details of the intranet router, the LAN configuration, and the SNA gateway.

7.4.2 Case Study Background and Requirements

In the previous case study we mentioned that NIT had secured access to a mainframe-based reservation system as part of its European acquisition. In this

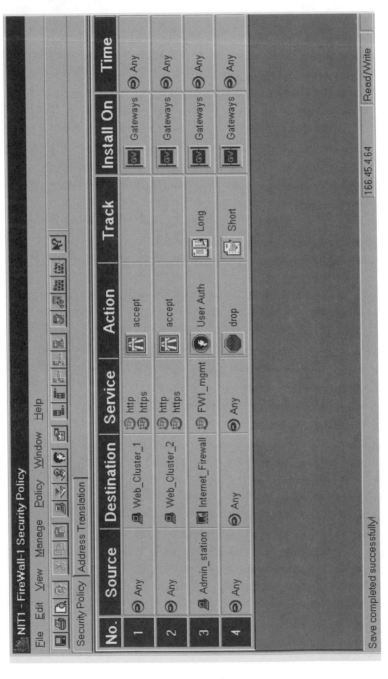

Figure 7.28 NIT security policy configuration using Firewall-1 policy administration GUI.

case study we will architect a solution that provides the United States–based operation access to this IBM mainframe system over a frame relay network. We will explore the various options for the communication protocol and the associated implementation options available in the market. In the end, we will provide a recommendation on the best option suited for NIT to complete the implementation details.

The requirements for this design can be stated as follows:

1. The NIT sales agents in the United States need to have access to the IBM mainframe in London using a TN3270 terminal-emulation client on their desktop.

2. NIT would like to make use of its current frame relay connectivity between New York and the London office (this connectivity was implemented in Case Study III).

3. The protocol of choice in the U.S. offices of NIT is TCP/IP and the LAN technology used is ethernet (IEEE 802.3).

Architecture

In a traditional IBM-only solution, the network architecture looked similar to Figure 7.29.

The IBM 3174 cluster controllers are the aggregation point for the TN3270 terminal and printer traffic. The aggregated traffic is then sent to the mainframe FEP via a dial-up, leased line connection using SDLC protocol. In a LAN environment, the cluster controller is connected to the FEP using a token ring. Over a period of time many changes have occurred. Network managers have found the technologies to replace the old IBM mainframe hardware. For example, the IBM 3270 dumb terminals have been replaced by TN3270 emulators running on PCs. The cluster controllers have been replaced by SNA gateways that talk TCP/IP to PCs and SNA to the mainframe. The controller-attached printers are replaced with LAN printers. On the WAN side, the slow private line or dial-up connections have been replaced with multiple, high-speed frame relay PVCs that carry LAN and SNA traffic side-by-side between the regional offices and the mainframe site. Another WAN alternative would be the public Internet, although the Internet-based access has limitation due to the strict timing requirements of SNA traffic. Frame PVCs can carry native SNA traffic using methods like RFC 1490 encapsulation or the data link switching (DLSw) tunneling. DLSw tunnels SNA traffic within TCP/IP. We will take a detailed look at each of these techniques and their pros and cons later in this case study. A sample logical setup with modern components is shown in Figure 7.30.

This frame-based architecture produces immediate cost savings due to:

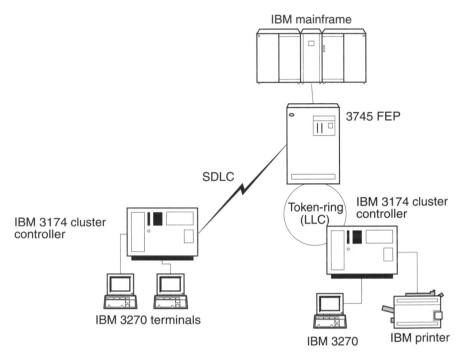

Figure 7.29 IBM-centric SNA connectivity to the mainframe.

1. Elimination of a long-haul private line for SDLC traffic;
2. Lower administration cost due to consolidation of LAN and WAN traffic over a single shared frame circuit;
3. Improved efficiency due to high-bandwidth mainframe connection;
4. Shared LAN printer for both mainframe and TCP/IP applications (this generic printer also reduces the cost of providing the print service);
5. Powerful gateway server based on a PC platform (thus it could be used as a shared server to run other applications as well as being easy to upgrade and maintain).

Transporting SNA Traffic Over the WAN

It is now time to take a detailed look at transporting the SNA traffic over a frame relay-based WAN. As we have mentioned before, there are primarily two methods to transport the SNA traffic over frame, RFC 1490 and data link switching. DLSw has been documented in detail in RFC 1795. Data link switching is a technique for transporting SNA traffic by tunneling through an IP network. The IP network could be using frame relay or the Internet as the

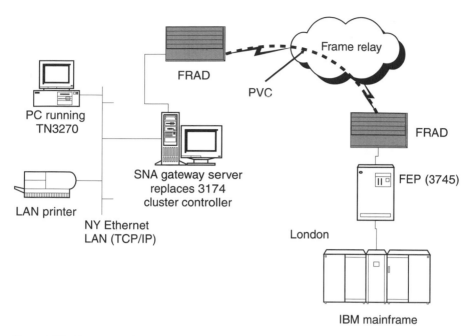

Figure 7.30 Network architecture for SNA connectivity using frame relay.

transport medium. The end routers accomplish the tunneling. The end routers in Figure 7.30 are the frame relay access devices, also known as FRADs. The DLSw terminates the data link control connections locally, thus avoiding the problem of link layer timeouts over the WAN. This is also known as local "spoofing" of the DLC connection. DLSw is primarily used in an environment that is IP centric and has a small percentage of SNA traffic. This is an ideal setup for IP-trained network managers. The IP tunneling introduces extra overhead per packet in the DLSw solution. When DLSw is transported over an IP backbone, the typical overhead is 56 bytes. If DLSw is being transported over an IP over frame, the overhead increases to 66 bytes. This is a lot of overhead compared to RFC 1490 encapsulation.

How does RFC 1490 work? RFC 1490 describes how to transport multiple protocols over frame relay. These protocols include IP, X.25, SNA, and others, while the DLSw is limited to encapsulation of SNA, APPN, and Netbios traffic only. Due to its multiprotocol nature, RFC 1490 is a popular method to transport various types of LAN traffic over the company intranet connected by a frame relay network. Many router vendors have implemented RFC 1490 in their routers, thus providing an interoperable infrastructure. Simply stated, RFC 1490 specifies how to encapsulate native SNA message units within a frame on a frame relay connection. This native encapsulation results in less overhead and an efficient use of the frame relay bandwidth. The RFC 1490 overhead consists of 16 bytes. This includes the frame flags, 2-octet address field, frame check sequence (FCS), type, control, and LLC2 fields. This overhead is only 25% of the DLSw overhead. Naturally, this saving translates into savings in network charges due to better efficiency and better application response time. RFC 1490 is used mainly in an SNA-centric environment as a direct replacement for the private long-haul SDLC circuits.

In conclusion of this discussion, if the WAN transport is based on frame relay, then RFC 1490 provides the most efficient method for transporting SNA data and other LAN data. However, if the WAN transport is based on an IP backbone (over private lines or the Internet), DLSw would be the encapsulation method of choice.

NIT has an additional choice in implementing the mainframe connectivity for its sales agents. This method is based on the popular browser front-end and has inherent cost savings due to a universal client. Advances in Web technology have produced a new alternative for secure mainframe access technology. The advent of universal, easy-to-use client and Java technology provides access to the mainframe and mainframe-based legacy applications without requiring any modifications to the legacy applications. Companies like Open Connect (www.oc.com) have developed Java-based applets for the TN3270 terminal emulation that can be downloaded to the client's browser, thus making use of a universal client interface that can be managed centrally and allows users to access the mainframe over the company's intranet or the Internet. This arrangement translates into savings in cost of licenses for the emulation software on each desktop (the browser in most cases is free), plus the yearly maintenance and upgrade cost. This concept is also well suited for the "thin-client" approach with Java stations or the network computers. Browser-based user access removes the limitation on the type of client platform, since the browsers are available for almost all platforms, and as long as the browser is Java enabled it will support the execution of a Java applet. This universal interface has a significant impact on the cost of providing user help desk and user training services. The Web browser-based solution is illustrated in Figure 7.31.

The server in front of the mainframe runs the Web server and the SNA gateway software. The client downloads a Java applet from the Web server and runs it within the browser environment. The Java applet sets up a persistent TCP/IP connection to the gateway server and runs the TN3270 protocol datastream from the client to the TN3270 server running on the SNA gateway server. The applet provides interpretation of 3270 datastreams and mapping of the TN3270 keyboard to the PC keyboard. One caution: this emulator applet may not support all the functions provided by a real 3270 terminal.

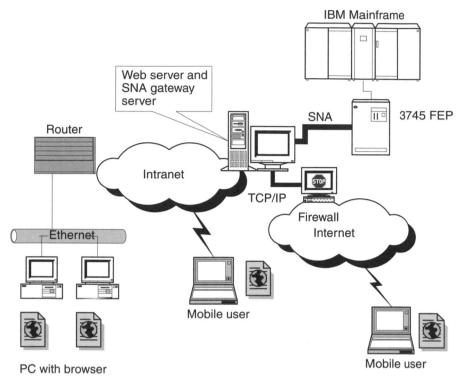

Figure 7.31 Browser-based access to the mainframe.

Security

How can one protect the mainframe in this new paradigm? Some of the key precautions would be to process all Internet access through a firewall that requires a session authentication. The authentication could be conducted at two places, one at the firewall and the second at the Web server. The authentication scheme could be based on a simple userID/password combination or a more complex token-based authentication. We have already looked at some of the token-based authentication schemes using the SecurID and CryptoCard. The Web server could be a simple Web server running on port 80 or a secure Web server. The advantage of a secure Web server is that the client can communicate with the server using a SSL protocol. The SSL technology provides encryption of the datastream, plus authentication of the server ID to the client, providing the necessary assurance to the client that it is communicating with the right server. Please refer to Chapter 5 on security for more details on SSL technology.

Implementation Details

In this case study, we will implement the architecture presented in Figure 7.32 using a SNA gateway server in New York. We will be using the routers in the previous case study. The FRAD1 and FRAD2 correspond to router 1 and router 2 in Case Study III. We will take a detailed look at each of these components.

Router R1 (FRAD1)

Router R1 has one frame circuit each on its serial interface 0 and serial interface 1. In a previous case study we had configured the secondary frame circuit on the subinterface serial 1.1, while a PVC to NOC was configured on the subinterface serial 1.1. For mainframe traffic we will configure separate PVCs on a new subinterface serial 1.3. The ethernet interface e1 has been configured on the sales and customer service subnet. The configuration for router R1 is given below (Table 7.18).

We will now take a close look at these configuration statements.

Frame relay map ip 166.42.44.18 28 broadcast ietf

This command statically maps the DLCI 28 to the destination IP address at the other end of the PVC. The configuration of the encapsulation type IETF generates RFC 1490 traffic. The RFC 1490 encapsulates SNA traffic into frames. Since the SNA traffic is time sensitive we want to give specific priority

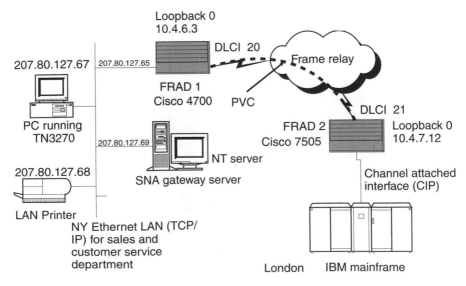

Figure 7.32 Detail architecture for access to mainframe over frame relay.

Table 7.18
Configuration of FRAD1 (Router R1)

---

1. *interface Serial1*

1. *description Secondary Frame relay connection from NY to London*

2. *no ip address*

3. *encapsulation frame-relay*

4. *frame-relay traffic-shaping*

5. *!*

6. *interface Serial1.1 point-to-point*

7. *description Secondary PVC over Link2 to London*

8. *ip address 166.42.44.9 255.255.255.252*

9. *frame-relay interface-dlci 24*

10. *!*

11. *interface Serial1.2 point-to-point*

12. *description Frame PVC to NOC*

13. *ip address 166.42.44.13 255.255.255.252*

14. *frame-relay interface-dlci 26*

15. *!*

16. *interface Serial1.3 point-to-point*

17. *description PVC for the mainframe traffic*

18. *frame-relay map ip 166.42.44.18 28 broadcast ietf*
Statically maps the DLCI to the destination IP and encapsulation type IETF generates the RFC1490 traffic.

19. *frame-relay class fast_vcs*   **## Traffic shaping on the PVC**

20. *!*

21. *map-class frame-relay fast_vcs*

22. *frame-relay traffic-rate 9600 16000*   **## Set the average and peak CIR value**

23. *frame-relay priority-group 2*

24. *!*

25. *priority-list 2 protocol sna high*   **## Set priority for certain class of traffic**

26. *priority-list 2 ip normal*

27. *!*

28. *interface Ethernet0*   **## This sets the ethernet interface**

29. *description Connection to the admin subnet*

30. *ip address 207.80.127.36 255.255.255.224*
Sets the interface IP address with the mask.

31. *ip broadcast-address 207.80.127.63* ## The mask is a VLSM creating 32 host subnet.

Table 7.18 *(continued)*

32.	*ip access-group 131 out*	**## Sets the ACL on this interface**
33.	*no ip redirects*	
34.	*no ip proxy-arp*	
35.	*no keepalive*	
36.	*!*	
37.	*interface Ethernet1*	
38.	*description Sales and Customer service subnet*	
39.	*ip address 207.80.127.71 255.255.255.224*	
		## Sets the interface IP address with the mask.
40.	*ip broadcast-address 207.80.127.95*	**## The mask is a VLSM creating 32 host subnet.**
41.	*ip access-group 132 out*	**## Sets the ACL on this interface**
42.	*no ip redirects*	
43.	*no ip proxy-arp*	
44.	*no keepalive*	

to the SNA traffic. This is important when many PVCs are sharing the frame circuit. In our configuration this is not as important since we have set up a dedicated PVC for the SNA traffic. The configuration is given here for completeness. The following command defines a new class called "fast_vcs" and then configures the average and peak CIR value on this PVC.

> *frame relay class fast_vcs*
> *map-class frame relay fast_vcs*
> *frame relay traffic-rate 9600 16000*

We then prioritize the traffic type and identify which packets are given precedence over others. The following commands prioritize SNA traffic over IP traffic:

> *frame relay priority-group 2*
> *priority-list 2 protocol sna high*
> *priority-list 2 ip normal*

Router R2 (FRAD2)

FRAD2 is a router R2 from case study three and is a Cisco 7505 router. We have added a new interface card to R2 known as the channel interface processor (CIP). The CIP provides a maximum of two channel-attached interfaces for the Cisco 7505 router, eliminating the need for a separate IBM FEP. CIP offers a combination of parallel channel adapter (PCA) and ESCON channel adapter (ECA). The PCA adapter also supports the "bus and tag" specifications, while the ESCON adapter supports the ESCON specifications. Mainframe channel is an intelligent processor that manages the communication with input/output (I/O) for the CPU, thus reliving the CPU cycles and providing concurrent operation of data processing and I/O.

The CIP consists of a motherboard and two adapter interface slots. For our configuration NIT needs to order a single ECA adapter on a dual ECA carrier (part number: CX-CIP-ECA1). The ECA has a female full-duplex connector. The ECA provides serial data transmission speeds up to 17 MB per second and signaling speed of 200 Mbps. The CIP can be attached to the mainframe channel using a fiber-optic ESCON cable. The CIP card insertion and the mainframe connection process should be carried out by shutting down the router. Since this might disrupt the current traffic, the activity should be carried out during a maintenance window. The CIP needs to be configured for the SNA (CSNA) environment. The CSNA feature supports SNA protocol over the ESCON interface to the mainframe. The CSNA support allows for communication between a channel-attached mainframe running VTAM and a LAN/WAN attached PU2.1 SNA node, which is our gateway server in New York.

To configure the CIP the following configuration steps are needed:

1. Load the appropriate microcode on the CIP interface card;
2. Configure the channel information;
3. Configure the internal LAN interfaces;
4. Configure bridging;
5. Configure the internal adapters link configuration.

Before we can configure the CIP interface we need to load the appropriate microcode on the interface from a flash card:

R2 (Config) # Configure microcode cip flash slot0:cipxxx-yy
R2 (Config) # Microcode reload
R2 (Config) #^Z   ## **Exit the configuration mode**
R2# show controllers cbus   ## **Display images loaded on the CIP card**
The "*cipxxx-yy*" is the microcode file.
Now we will configure the CIP for CSNA support.

R2 (Config) #interface channel3/0   ## **The CIP card has been inserted in slot 3 on the router and the "0"**

refers to the ESCON adapter position on the CIP card

R2 (Config-if) # csna 0200 45   ## These are the path and device values
for the channel

Configuring the internal LAN interfaces on the CIP interface card.

R2 (Config) #interface channel3/2   ## The CIP card has been inserted
in slot 3 on the router and the "0"
refers to the internal LAN adapter
position on the CIP card
R2 (Config-if) # max-llc2-sessions1024   ## Defines the maximum concurrent LLC2 sessions

Configuring bridging for the internal ethernet LAN.

R2 (Config-if) # lan ethernet 0   ## Enter the LAN configuration mode
R2 (Config-lan) # bridge-group 1   ## Assign a bridge-group

Configure the adapter for the internal ethernet LAN.

R2 (Config-lan) # adapter 0 6000.0001.06A2   ## configures the adapter
number and an MC ad-
dress

Now you can configure the link characteristics like idle-time, ack-delay
and retry count.

Table 7.19 provides the configuration for router R2.

7.4.3   Conclusion

This concludes our implementation of NIT's mainframe access. In this case
study we investigated various options for the mainframe access and implemented a combination of SNA gateway server and Cisco Channel Attached
Interface solution. Cisco also offers an SNA server implementation on the CIP
card. This eliminates the SNA gateway server requirement in New York and
SNA-encapsulated IP traffic over the frame relay connection. In this scenario,
the emulation clients can make a TCP/IP connection to the Cisco router directly
over the WAN. The CIP will then switch this connection using the internal ethernet LAN to the mainframe channel. Although this is much straightforward
implementation, it has a disadvantage of putting a heavy load on the router since
all the packets are process-switched by the router.

Table 7.19
Configuration of FRAD2 (Router R2)

```
!
interface Serial1/1
description Secondary Frame relay connection from London to NY
 no ip address
 encapsulation frame-relay
!
interface Serial1/1.1 point-to-point
 description Secondary PVC over Link2 to London
 ip address 166.42.44.10 255.255.255.252
 frame-relay interface-dlci 25
!
interface Serial1/1.2 point-to-point
 description PVC for the mainframe traffic
frame-relay map ip 166.42.44.17 29 broadcast ietf
                         ## Statically maps the DLCI to the destination IP
                         ## Encapsulation type IETF generates the RFC140 traffic
!
interface channel3/0
csna 0200 45
!
interface channel3/2
max-llc2-sessions1024
!
lan ethernet 0
bridge-group 1
adapter 0 6000.0001.06A2
!
no keepalive
```

7.5 Case Study V: Intranet Connectivity in a Company Using Internet-Based VPN

7.5.1 Case Study Objective

At the end of this case study you will be able to set up an intranet within a company with offices in multiple geographical locations. In the third case study, we implemented the intranet connectivity using frame relay technology. In this case study we will implement the intranet connectivity using VPN technology. As we have seen in Chapter 6, VPN technology allows us to set up secure and cost-effective private connections between any two locations connected to the public Internet. We will implement VPN-based intranet connectivity between NIT's branch offices and HQ using a software-based VPN solution. The software solution used in this case study is Checkpoint's Firewall-1 software running on a Sun Solaris operating system.

7.5.2   Case Study Background and Requirements

NIT has acquired another travel services company in the United States, with offices in Los Angeles and Seattle. Both of the offices are connected to the Internet using a fractional T1 connection of 512 Kbps. NIT currently has a 256 Kbps connection to the Internet. NIT's Internet connection was set up with a Checkpoint firewall in Case Study III. Due to the presence of the Internet connections at the new offices, NIT would like to connect the new offices to its corporate headquarters in New York with a VPN connection over the public Internet. NIT would also like to give its mobile sales force a new secure dial-access mechanism to access the company intranet over the Internet. The sales force can now use MCI's local dial numbers throughout the United States. This will allow NIT significant cost savings, since it can add dial access to the intranet without additional investment in its own modem pool. In this case study we will set up the LAN-to-LAN VPN, while the next case study will focus on the secure dial-in access to the intranet and extranet.

Architecture

The advance in cryptography and authentication services allows us to set up secure IP "tunnels" or VPN over the public Internet. We will make use of various authentication and encryption schemes discussed in security-related topics in Chapter 5 to set up these secure IP tunnels. In this case study we will set up a LAN-to-LAN VPN based on Checkpoint's software-based solution.

Checkpoint's Software-Based Solution Using Firewall-1 Software

In order to set up Firewall-1 to a Firewall-1 VPN, we need a Checkpoint Firewall-1 installation at the Seattle and Los Angeles offices. If we have IPSec-compliant firewalls at these locations, we can set up IPSec tunnels with the Checkpoint firewall at the New York site. Since the new sites currently do not have firewalls we will deploy Checkpoint Firewall-1 solution at these sites. This will allow NIT to deploy one vendor firewall solution across the intranet. This has a natural cost advantage and operational simplicity, since NIT can fully utilize its Firewall-1 trained resources and use one security policy management platform across the intranet.

The architecture with a Firewall-1–based VPN is shown in Figure 7.33.

We have already seen the basic Firewall-1 installation steps in Case Study III. The key difference in the configuration of FW-NY (Firewall at New York) is the remote module configuration section. In the previous configuration (Case Study III) of FW-NY we had followed the configuration step below during the initial installation.

```
Configuring Remote Modules
================================
Remote modules are FireWall or Inspection modules that are
going to be controlled by this Management Station.
Do you want to add Remote modules (y/n) [y] ? n
```

In this case study we will reconfigure the FW-NY to be the management station for the remote firewalls, FW-SE (Firewall at Seattle) and FW-LA (Firewall at Los Angeles) as shown in Figure 7.33. This allows the firewall administrators in New York full control over the remote firewalls and enables them to implement security policies from a single management station.

Note A management station is a workstation on which the Firewall-1 security policy is defined and maintained.

Follow the steps below for Firewall-1 installation on FW-SE and FW-LA. The key difference between these installations and the installation of the FW-NY

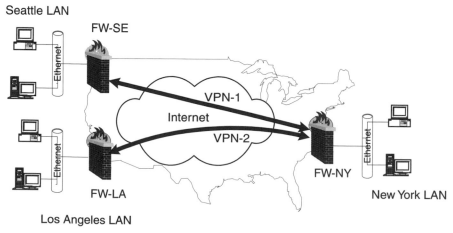

Figure 7.33 Logical VPN architecture.

in Case Study III is that here we will only install the firewall module. The FW-NY has both the firewall module and the management module. The management module in FW-NY will be used to manage the firewall policies on FW-SE and FW-LA.

```
# ./fwinstall ↵
************** FireWall-1 v3.0 Installation **************
Reading fwinstall configuration.  This might take a while.
Please wait.
Configuration loaded.  Running FireWall-1 Setup.
Please read the following license agreement.
Hit 'ENTER' to continue... ↵
        SOFTWARE RIGHT-TO-USE LICENSE AGREEMENT
( ## The software licensing terms and conditions go here.)
Do you accept all the terms of this license agreement (y/n)
? y
Checking available options. Please wait...................
Which of the following FireWall-1 options do you wish to
install/configure ?
-----------------------------------------------------------
(1) FireWall-1 Enterprise Product
(2) FireWall-1 Single Gateway Product
(3) FireWall-1 Enterprise Management Console Product
(4) FireWall-1 FireWall Module
(5) FireWall-1 Inspection Module
Enter your selection (1-5/a): 4 ↵
Installing/Configuring FireWall-1 FireWall Module.
Which FireWall Module would you like to install ?
----------------------
(1) FireWall Module/25
(2) FireWall Module/50
(3) FireWall Module (Unlimited)
Enter your selection (1-3/a) [3]: ↵
## The default option is shown in the brackets
Please wait...
Selecting where to install FireWall-1
--------------------
FireWall-1 requires approximately 10429 KB of free disk
space.
Additional space is recommended for logging information.
Enter destination directory [/etc/fw]): ↵
```

```
Checking disk space availability...
Installing FW under /etc/fw (344034 KB free)
Are you sure (y/n) [y] ? ↵


Software distribution extraction
----------------
Creating directory /etc/fw
Extracting software distribution. Please wait ...
Software Distribution Extracted to /etc/fw
Installing license
--------------
Reading pre-installed license file fw.LICENSE... done.


The following evaluation License key is provided with this
FireWall-1 distribution
Eval           15Oct97    3.x controlx pfmx routers con-
nect motif embedded.


Do you want to use this evaluation FW-1 license (y/n) [y]?
↵
## Choose this license for now, we will add the correct li-
cense later with 'fw putlic' command.
Using Evaluation License String
This is FireWall-1 Version 3.0b [VPN] ( 6Aug98 17:40:15)
Type          Expiration Ver Features
Eval          15Oct97    3.x controlx pfmx routers connect
motif embedded
License file updated
Module /etc/fw/modules/fwmod.5.3.o was updated.


********* FireWall-1 kernel module installation **********
installing FW-1 kernel module...
Done.
Do you wish to start FireWall-1 automatically from
/etc/rc3.d (y/n) [y] ? ↵


FireWall-1 startup code installed in /etc/rc3.d
Welcome to FireWall-1 Configuration Program
=============================================
This program will guide you through several steps where you
will define your FireWall-1 configuration. At any later
```

time, you can reconfigure these parameters by running fw-config
Configuring Licenses...
============--=========
The following licenses are installed on this host:
Eval   15Oct97   3.x controlx pfmx routers con-
nect motif embedded

Do you want to add licenses (y/n) [n] ? ↵
Module /etc/fw/modules/fwmod.5.5.1.o was updated

Configuring Masters...
======================
Masters are trusted Management Stations which are going to
control this FireWall Module.
Do you want to add Management Stations (y/n) [y] ? ↵
Please enter the list hosts that will be Management
Stations.
Enter hostname or IP address, one per line, terminating
with CTRL-D or your EOF character.
204.80.127.2   ## IP address of FW-NY
^D
Is this correct (y/n) [y] ? ↵
You will now be prompted to enter a secret key that will
be used to authenticate and encrypt the communication be-
tween this Module and the Management Stations that you have
selected.
Enter secret key: xxxxx
Again secret key: xxxxx

NOTE:
Do not forget to run 'fw putkey' with the same secret key
on each of the configured masters.
Configuring Security Servers...
===============================
Installing Secured Services option

You may make the services below secured. By doing that you
will enable the usage of strong authentication and/or content security for this service (see the user guide for details).

```
1) FTP
2) HTTP
3) TELNET
4) RLOGIN
5) SMTP
```

Please enter the numbers of the services you want to make secured.

For example, if you wish to make only ftp and rlogin secured,

enter '1 4' > 4

You have selected rlogin to be secured.

Is this correct (y/n) [n]: y

Do you wish to enable the client authentication feature (y/n) [y] ? n

In version 2.1, the FTP, Telnet and HTTP security daemons were listening on their original TCP ports (i.e. 21, 23 and 80 respectively). Beginning at version 3.0, these daemons may be installed on randomly selected high ports instead. The FireWall gateway will redirect connections coming to the original ports to these high ports.

Answering 'yes' to the following question will keep the FTP, Telnet and HTTP security daemons on their original TCP ports. This will allow security policies to be loaded from version 2.1 control module on a version 3.0 gateway module.

Do you wish to enable backward compatibility, i.e. let 2.1 control stations control 3.0 inspection modules (y/n) [n] ? ↵

Restarting the inet daemon (process 124)
Configuring Groups...
======================
FireWall-1 access and execution permissions

Usually, FireWall-1 is given group permission for access and execution. You may now name such a group or instruct the installation procedure to give no group permissions to FireWall-1. In the latter case, only the Super-User will be able to access and execute FireWall-1.

Please specify group name [<RET> for no group permissions]: fwgroup

```
Group staff will be used. Is this ok (y/n) [y] ? ↵
Setting Group Permissions... Done.
Configuring IP Forwarding...
_=========___ ˉ==========----==
Do you wish to disable IP-Forwarding on boot time (y/n) [y]
? ↵
IP forwarding disabled
Configuring Default Filter...
==============================
Do you wish to modify your /etc/rcS.d boot scripts to allow
a default filter to be automatically installed during boot
(y/n) [y] ? ↵
Which default filter do you wish to use?
----------------------------
(1) Allow only traffic necessary for boot
(2) Drop all traffic
Enter your selection (1-2) [1]: ↵
Generating default filter
Configuration ended successfully
************** FireWall-1 is now installed. **************
Do you wish to start FW-1 now (y/n) [y] ? n
*********************************************************
               DO NOT FORGET TO:
1. add the line:    setenv FWDIR /etc/fw    to .cshrc
          or  FWDIR=/etc/fw; export FWDIR to .profile
2. add  /etc/fw/bin  to path
3. add  /etc/fw/man  to MANPATH environment
*********************************************************
You may configure FireWall-1 anytime, by running fwconfig.
*********** Installation completed successfully **********
```

We will now add the correct license using the "fw putlic" command. The license is generated by Checkpoint licensing center and is specific to the hostID on Unix workstations. For example, 72761388 is the host ID of FW-LA. The license key and then the feature sets that we have bought for this firewall follow the hostID. This includes the encryption algorithms.

```
     # fw putlic 0x72761388 35c9d1e8-eb809371-ff730457 pfmx
des skip activemod controlx routers des skip motif embed-
ded
```

```
     This  is  FireWall-1  Version  3.0b  [VPN]  (  6Aug98
17:43:26)
     Type     Expiration Ver Features
     ID-72761388    6Aug98       3.x controlx pfmx routers
des skip connect motif embedded
     Eval     15Oct97      3.x controlx pfmx routers connect
motif embedded
     License file updated
     Module /etc/fw/bin/../modules/fwmod.5.5.1.o was  up-
dated
     # fwstart
     FW-1: Starting fwd
     FW-1: Starting snmpd
     snmpd: Opening port(s): 161
     SNMPD: server running
     FW-1: Fetching Security Policy from 204.70.127.2 lo-
calhost
     Trying to fetch Security Policy from 204.70.127.2:
     ## The firewall module fetches the security policy in
order  specified  in  the  '/etc/fw/conf/masters'  file.  The
management servers  are  specified  in  the  order  of  prefer-
ence. If the firewall module is  unsuccessful in retrieving
a security policy from the configured masters, local host
is accessed for a security policy as a last resort.
FW-1 started
#
After the installation of FW-SE and FW-LA configure FW-NY
to add the remote firewalls to its management domain. This
creates a 'master-slave (client)' relationship between FW-
NY and the new firewalls. Following steps illustrate the
configuration process.
# fwconfig ↵Welcome to FireWall-1 Configuration Program
=============================================
This program will let you re-configure your FireWall-1 con-
figuration.
Configuration Options:
------------
(1)   Licenses
(2)   Administrators
(3)   GUI clients
(4)   Remote Modules
```

```
(5)   Security Servers
(6)   Groups
(7)   IP Forwarding
(8)   Default Filter
(9)   CA Keys
(10)  Exit


Enter your choice (1-10) :4 ↵


Configuring Remote Modules...
================================
Remote Modules are FireWall or Inspection Modules that are
going to be controlled by this Management Station.
Do you want to add Remote Modules (y/n) [y] ? ↵
Please enter the list hosts that will be Remote Modules.
Enter hostname or IP address, one per line, terminating
with CTRL-D or your EOF character.
207.80.131.1  # IP address of FW-SE on the Internet con-
nected Interface
207.80.132.1  # IP address of FW-LA on the Internet con-
nected Interface
^D
Is this correct (y/n) [y] ? ↵You will now be prompted to
enter a secret key that will be used to authenticate and
encrypt the communication between this Management Station
and the Remote Modules that you have selected.
Enter secret key: xxxxxx ↵
Again secret key: xxxxxx ↵
```

Note Do not forget to run "fw putkey" with the same secret key on each of the configured clients. The communication between the master and client is validated using this secret key, also known as the firewall authentication password.

```
Log into FW-SE and FW-LA and run following command
fw putkey fw-NY <key> ↵
Remote Modules Configured successfully
Configuration Options:
-----------------
(1)   Licenses
(2)   Administrators
(3)   GUI clients
```

```
(4)    Remote Modules
(5)    Security Servers
(6)    Groups
(7)    IP Forwarding
(8)    Default Filter
(9)    CA Keys
(10) Exit
Enter your choice (1-10) :10 ↵
Thank You...
#
```

Now we are ready to configure the VPN setup between FW-NY-to-FW-SE (VPN-1) and FW-NY-to-FW-LA (VPN-2) as shown in Figure 7.33. The key to the VPN setup is the proper selection of encryption and authentication algorithms that provide privacy, authenticity, and data integrity while the data is traversing the public Internet. Firewall-1 supports three encryption schemes:

1. FWZ, Checkpoint's proprietary encryption scheme;
2. Manual IPSec, IETF-specified standard encryption and authentication scheme with manual key management;
3. SKIP, simple key management for IP from Sun Microsystems.

If all the firewalls in an intranet are implemented based on Checkpoint Firewall-1 software, we can very easily set up the VPN using the FWZ encryption scheme. FWZ uses a MD5 algorithm for authentication and DES or FWZ1 algorithm for message encryption. The encryption key is derived using Diffie-Hellman algorithm. A CA present in Firewall-1 certifies the Diffie-Hellman keys.

If one needs interoperability, Manual IPSec is the best choice. Manual IPSec has its limitations due to scaling issues in distributing keys to all sites and a potential security threat due to fixed keys. SKIP has also seen a wide deployment due to Sun's early initiative in this area. For this case study we will use the FWZ encryption scheme.

Key steps to implement a FWZ-based VPN:

1. Define encryption domain for each firewall. The encryption domain includes network objects like communicating subnets and hosts behind a firewall.
2. Configure encryption scheme to be FWZ, on each firewall.
3. Set up and exchange Diffie-Hellman public keys among communicating firewalls for FWZ encryption scheme.

4. Configure VPN rules in the security policy on each firewall.

5. Configure FWZ encryption scheme for the VPN rule.

6. Install security policy on each firewall.

Let's take a detailed look at each step.

Define Encryption Domain for Each Firewall

The encryption domain includes network objects like communicating subnets and hosts behind a firewall. Please refer to Figure 7.34 for a definition of encryption domain for each site.

Figure 7.35 shows the configuration of FW-NY. The configuration defines the IP address of the firewall, its name, location, and type.

Next, click on the "Interface" tab to configure both the internal (le0) and external (le1) interface on the firewall. Figure 7.36 shows the configuration of the le0 interface, and Figure 7.37 shows the completed interface configuration with both interfaces shown in the interface properties window.

Figure 7.38 configures the various authentication schemes allowed while communicating with the firewall.

Configure Encryption Scheme to Be FWZ, on Each Firewall

One last step before we configure the encryption domain is to configure a group that identifies all the network objects in the encryption domain of the FW-NY. Use the security policy GUI to create a new group called "NY-encdomain" under network objects. Figure 7.39 shows the specification of NY-encdomain.

Now we are ready to configure the encryption domain and the encryption scheme used for FW-NY. Figure 7.40 shows this configuration.

Depending on your license agreement, you will have additional encryption methods like SKIP or Manual IPSec. In Figure 7.40 we have selected the group NY-encdomain and FWZ encryption method.

Set Up and Exchange Diffie-Hellman Public Keys Among Communicating Firewalls for the FWZ Encryption Scheme

Click on the Edit button under the "encryption method window" in Figure 7.40. This opens the FWZ properties window shown in Figure 7.41. Generate a CA public key for FW-NY. When you click on "Generate" an RSA key pair is generated and the public key portion is displayed in the window as shown in Figure 7.41.

Select the DH Key (Diffie-Hellman) tab, and generate the Diffie-Hellman key pair. The public key portion is shown in Figure 7.42.

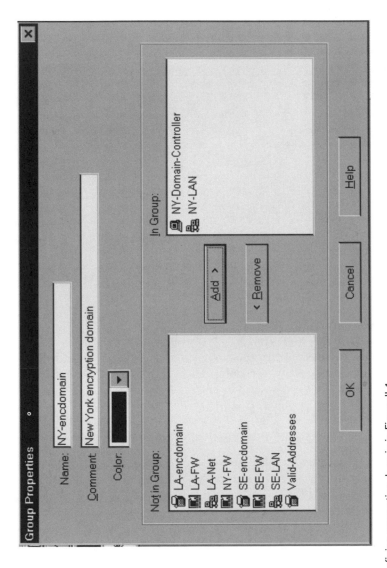

Figure 7.34 Defining encryption domain in Firewall-1.

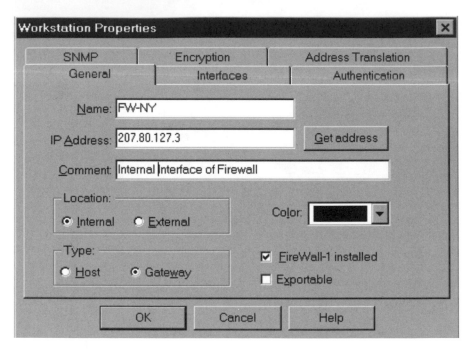

Figure 7.35 Defining the firewall at the New York City location.

Configure FW-SE and FW-LA following steps one through three. Now go back to the FWZ properties screen as shown in Figure 7.41, and select remote option and select any of the remote firewalls from the drop-down menu and get the remote firewall's CA public key. Click the DH Key tab and retrieve the remote firewall's DH public key. Now we are ready to configure the VPN rule on each firewall (see Figure 7.43).

Configure VPN Rules in the Security Policy on Each Firewall

The first rule specifies that any network traffic originating from and destined to the New York office LAN from the LAN in the Los Angeles office is to be encrypted. The second rule specifies a similar rule for traffic between the New York and Seattle LAN.

Configure FWZ Encryption Scheme for the VPN Rule

In the VPN rule we have specified "encrypt" action. We can specify the encryption scheme using edit property for the action. Position your pointing device on the Encrypt Action and click the right button. Select the Edit Properties menu from the drop-down menu. Upon selection, you will see the screen shown in Figure 7.44.

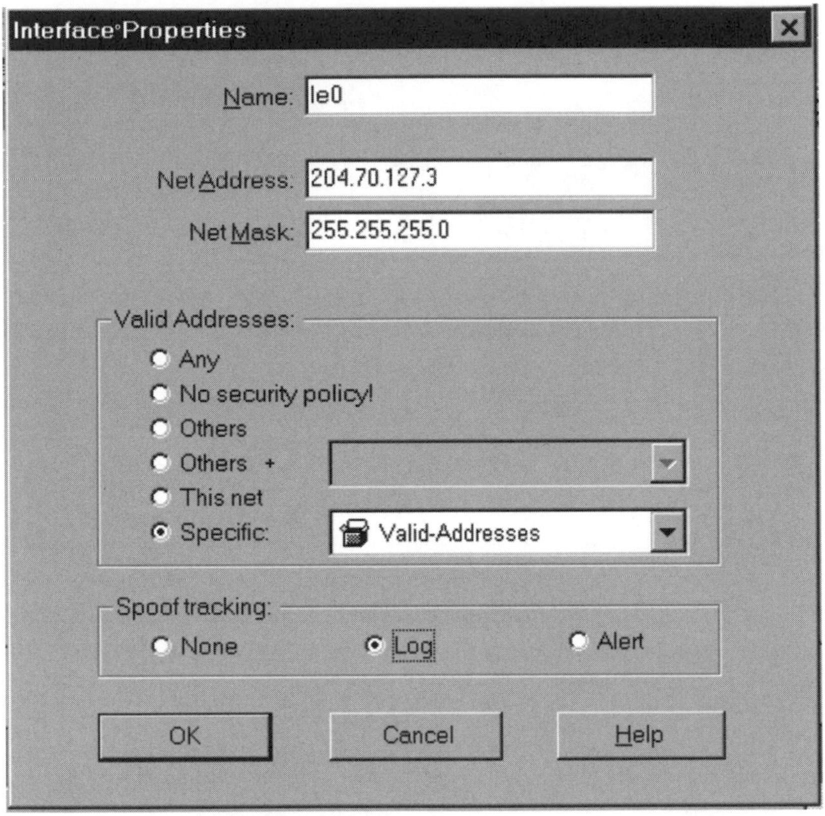

Figure 7.36 FW-NY interface properties.

If your license allows additional encryption schemes like SKIP and Manual IPSec, these options will appear in this window. Select FWZ scheme and then click on the Edit button to specify properties of this encryption scheme. Specify encryption methods for the session and data in the screen shown in Figure 7.45.

In the Allowed Peer Gateway property, you can specify a specific firewall or allow any firewall connection as permitted by the firewall policy defined in the policy rule. The session and data encryption methods could be FWZ1 (Checkpoint's proprietary scheme), DES, 3DES, or none (i.e., clear transmission). Data integrity will be checked based on a MD5 checksum.

Install Security Policy on Each Firewall

The last step in this process is to install the firewall policy on the firewall. Verify and save the policy and install it on respective firewalls.

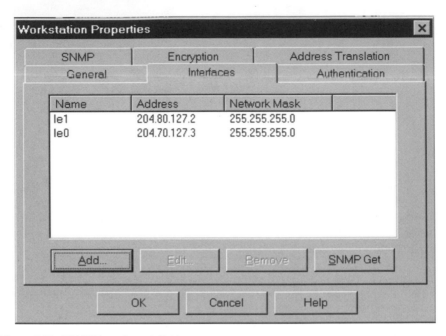

Figure 7.37 FW-NY interface definitions.

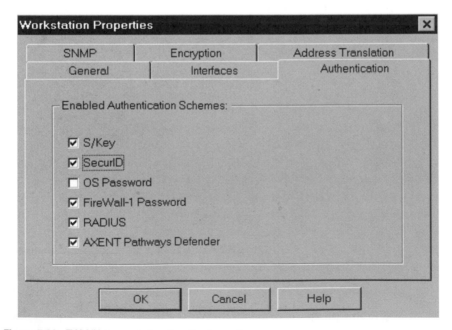

Figure 7.38 FW-NY supported authentication schemes.

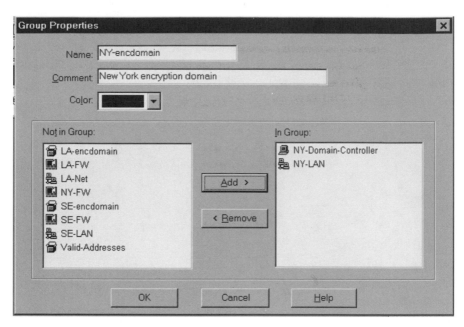

Figure 7.39 NY encryption domain definition.

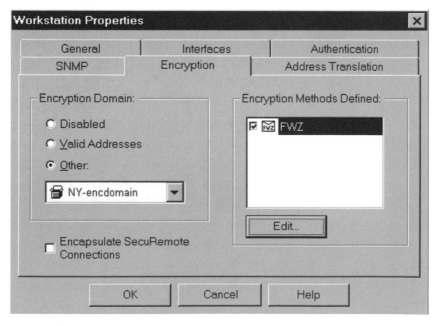

Figure 7.40 FW-NY encryption method.

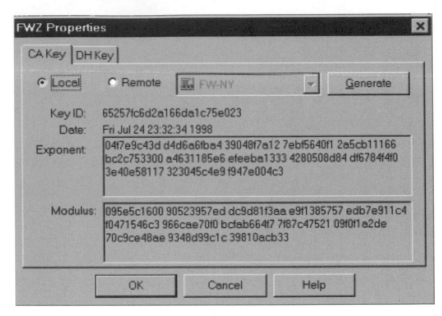

Figure 7.41 CA key for FWZ.

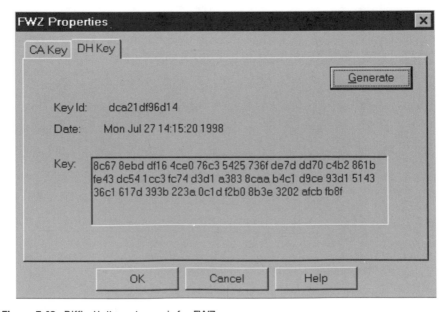

Figure 7.42 Diffie-Hellman key pair for FWZ.

Figure 7.43 VPN rule using the Security Policy Manager GUI.

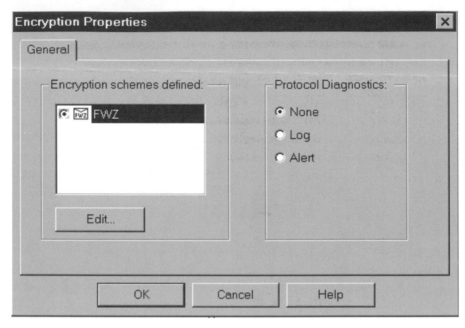

Figure 7.44 Encryption properties.

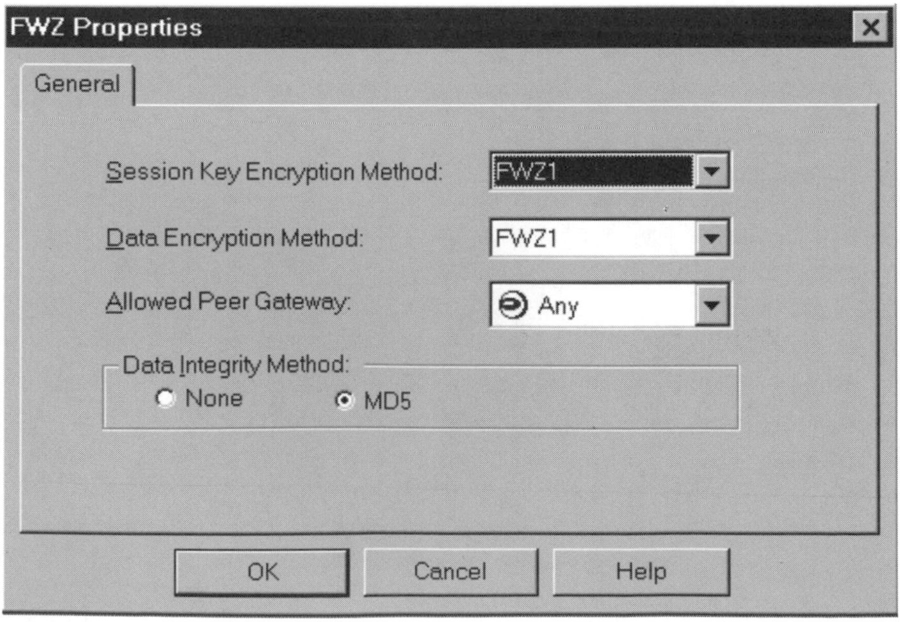

Figure 7.45 FWZ properties.

7.5.3 Conclusion

In this case study we have explored the set up of VPN over the public Internet using Checkpoint's Firewall-1 solution. We have set up the policy for VPN, the encryption method for the firewall, and the defined properties of the encryption method. We have learned to use Checkpoint's Security Policy Manager GUI to create and install a security policy on a Firewall-1 server.

7.6 Case Study VI: Remote Access to the Intranet Using Internet-Based VPN

7.6.1 Case Study Objective

At the end of this case study you will be able to set up an intranet connection for a mobile user over the Internet using solutions from Checkpoint and Cisco. We will also look at various options for such a setup and then evaluate the pros and cons of each option.

7.6.2 Case Study Background and Requirements

NIT would like to offer its employees remote dial-in access to its corporate intranet. The intranet remote access will initially be for the New York area followed by Seattle and then Los Angeles. After the successful implementation in the United States, NIT would like to give dial-in remote access to its mobile users in Europe. The remote access will provide mobile and telecommuting employees access to the corporate LAN and intranet servers. The remote access setup must be easy, secure, and cost effective.

Architecture

The advances in cryptography and authentication services allow us to set up secure IP tunnels or VPNs over the public Internet (please refer to Chapter 6 for VPN discussion). We will make use of IPSec-based authentications and encryption schemes discussed in Chapter 6. This allows us to set up secure IP tunnels over the Internet. In this case study we will set up a mobile user-to-LAN VPN based on:

- Checkpoint Firewall and SecuRemote software;
- Cisco router and RedCreek's RavlinSoft Remote Access Client.

The remote access solutions assume a dial-up access either to the Internet or directly to the corporate LAN. Let's examine the options for setting such an infrastructure.

Dial-In Infrastructure

To set up a dial-in infrastructure for the remote access to the corporate LAN, NIT has two choices—to set up the dial-in infrastructure in-house or to outsource the dial-in infrastructure to a service provider like MCI WorldCom or GTE. The general rule of thumb is that if you anticipate a small number of dial-in users and you would like to maintain a strong control over the security, set up your own infrastructure. If you anticipate a rapid growth in the population of remote access users, you should seriously consider outsourcing the whole remote access effort to a service provider. The following are the pros and cons of each approach:

Pros of keeping the remote access infrastructure in-house:

- In some firms, a significant installation of dial-up ports already exists;
- Full control over the operational security;
- Can control the quality of service to certain extent;
- For a small operation there is a cost advantage.

Cons of keeping the remote access infrastructure in-house:

- In case of a rapid growth, the solution is not scalable.
- It is very costly to keep up with the changing and improving technology; this results in technological obsolescence.
- High cost of labor. An in-house solution eats into the IT budget for labor and maintenance equipment. For example, a company would have to maintain at least two network engineers (one backs up the other) and an operations staff for the customer help desk.
- The customer support may not be 7×24, 365 days a year.

By outsourcing the remote access service to a service provider a corporation can guard against all the cons listed for the in-house setup plus add the following pros:

1. Higher performance and reliability;
2. Better national and global coverage (this results in lower cost of long-distance calls);
3. Better security arrangements by the service provider;
4. The organization can focus its resources on its core business.

Figure 7.46 shows the dial-in architecture for an in-house setup.

The network access server (NAS) provides the modem pools for users to dial-in. In this architecture, NIT would need to get a connection to the PSTN from the local RBOC or a CLEC. The RBOC will provide the telephone numbers and the physical common business lines (CBL) to the NIT premises. In this architecture, the New York and Seattle sites maintain their own NAS servers. The primary access control server in New York performs the dial-up authentication. If the primary server does not respond then the backup access control server in Seattle provides authentication.

Alternatively, Figure 7.47 shows the architecture for the remote access infrastructure that has been outsourced to a service provider.

In order to meet the rapidly expanding needs of remote access, NIT has chosen to outsource the dial-in infrastructure to a national service provider. Thus, the service provider will provide the dial-up numbers and the access to the Internet while NIT will manage the VPN tunnels. In order to choose a national service provider, NIT issued a request for proposal. The service provider was chosen based on the responses to the RFP. The RFP focused on the following key issues:

- Current infrastructure of the service provider in the United States and specifically in the New York, Seattle, and Los Angles areas;

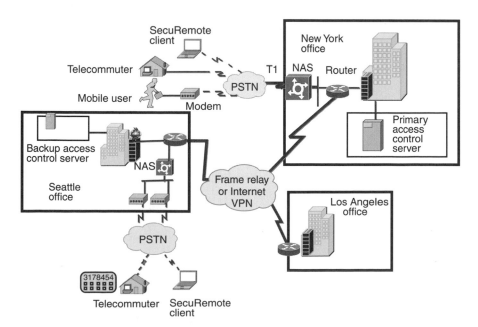

Figure 7.46 In-house dial-in architecture for remote access.

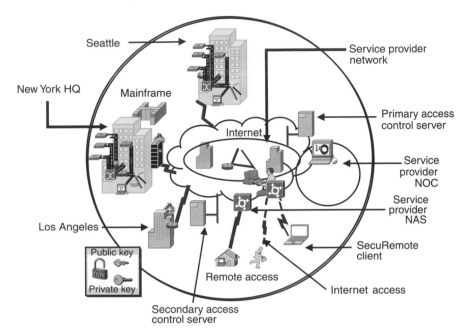

Figure 7.47 Remote access architecture using a service provider.

- Support for dial-in technologies like V.90 56k, ISDN, ADSL, and cable modem;
- Performance and availability of SLAs;
- Security infrastructure, specifically the technology used for authentication and encryption;
- Service provider arrangements with other ISPs for international dial-roaming access.

Now that the dial-up infrastructure has been outsourced to a service provider, let's implement the secure VPN tunnels over the Internet using Checkpoint and Cisco solutions.

Checkpoint's Software-Based Solution Using Firewall-1 Software

Due to the current installations of Checkpoint Firewall-1 solution at NIT, it is very easy to set up a SecuRemote-based VPN for a mobile client. This has a cost advantage and operational simplicity, since NIT can fully utilize its Firewall-1 trained resources and use one security policy management platform across the intranet.

The Checkpoint SecuRemote client software will be installed on the laptops and desktops of mobile and telecommuting users. The NAS from the service provider provides the modem pools for users to dial-in. We will now configure the SecuRemote client and the Checkpoint Firewall-1 software to set up a VPN tunnel across the Internet as shown in Figure 7.48.

The VPN is based on IPSec encryption and a RSA authentication scheme. Checkpoint offers other choices for encryption technology such as SKIP, FWZ (Checkpoint proprietary), or DES encryption. The SecuRemote software transparently encrypts all TCP/IP communication from the laptop network adapters. This includes the dial-up adapter as well as the local ethernet adapter. SecuRemote is available for Windows 95 and Windows NT 4.0. The encrypted and authenticated VPN session prevents any eavesdropping even if the data is traveling over the public Internet. Mobile users can dial-in on a local number provided by the service provider, thus saving money on long-distance charges to NIT. Customers of NIT can also use this mode to gain access to the NIT extranet in the future. The access controls on the Firewall-1 server can be configured at various levels of granularity, for intranet or extranet access.

Client-Side Configuration on Windows 95

1. Install the SecuRemote client software from the diskette. Make sure that you are running the Windows 95B/98 version of the OS.

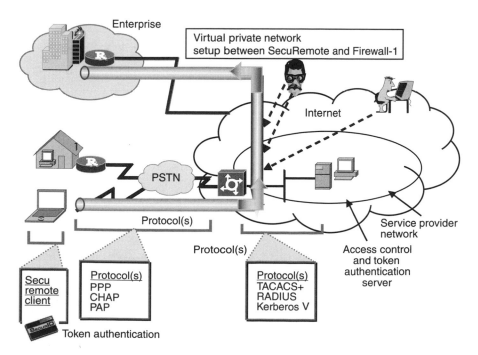

Figure 7.48 VPN tunnel using Checkpoint Firewall-1.

2. The installation process will install the "kernel" module that sits in the TCP/IP kernel, and a daemon that runs in the background. An icon for the daemon will appear in your system tray.

3. The kernel module monitors the incoming and outgoing TCP/IP traffic. If the traffic is destined to or coming from an encryption domain, the traffic is encrypted. The kernel module gets this information from the "daemon" process.

4. The daemon knows the addresses belonging to the encryption domain because the user configures these addresses during the setup. Once the kernel module has informed the daemon of an outgoing packet, the daemon will hold the first packet and determine the remote site and the remote firewall. The daemon initiates a Diffie-Hellman key exchange and then uses those keys for encrypting and authenticating the information using RSA algorithms.

5. The daemon process is configured with the information on remote sites by selecting "Make New Site" menu under "Sites" on the GUI. The daemon then queries the remote management servers for the RSA key, a complete list of firewalls, and their encryption domains. You can kill the daemon by selecting the "kill" option with a right mouse click.

6. The username/password for access through the firewall could be based on token authentication or a regular password. In case of a regular password, the userID/password combination can be preprogrammed in the daemon.

Server-Side Configuration

In order for the Firewall-1 server to accept the SecuRemote connections the following steps are needed:

1. Configure an encryption domain on the server side for the New York network. This step was outlined in detail in Case Study V, in context of the Firewall-1-to-Firewall-1 VPN solution. The "NY-Encryption" domain consists of the New York LAN and the hosts on that LAN. Make sure that the box "exportable" on the New York Firewall-1 server configuration is selected (see Figure 7.35). This allows the server to share the information about its encryption domain to all the requesting SecuRemote clients. Firewall-1 configuration at HQ for Internet VPN access is shown in Figure 7.49.

2. Make sure you have an encryption module installed during the Firewall-1 server installation.

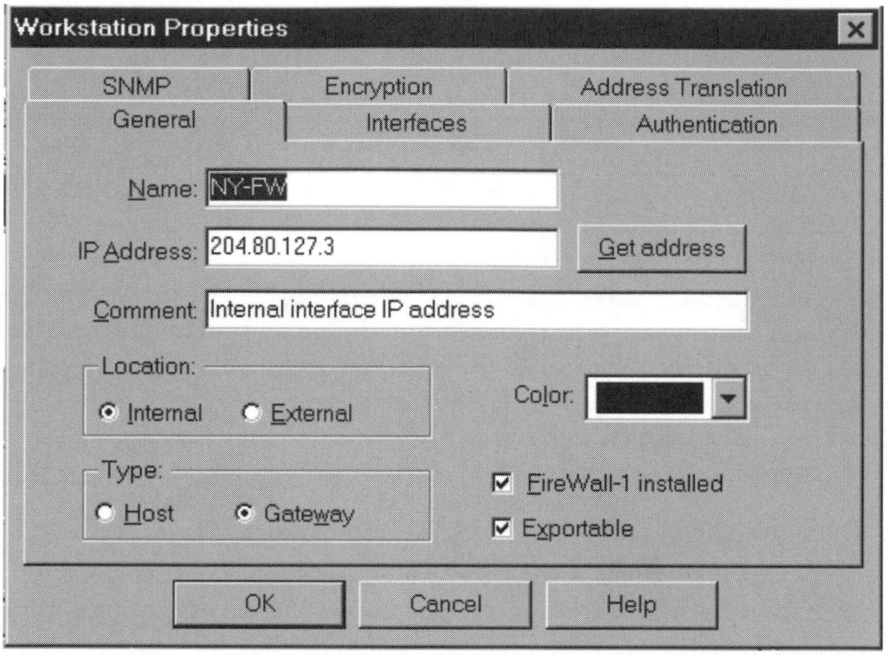

Workstation Properties

SNMP	Encryption	Address Translation
General	Interfaces	Authentication

Name: NY-FW

IP Address: 204.80.127.3   Get address

Comment: Internal interface IP address

Location:
  ◉ Internal   ◯ External

Color: ▬▬▬▬ ▼

Type:
  ◯ Host   ◉ Gateway

☑ FireWall-1 installed
☑ Exportable

[OK]   [Cancel]   [Help]

(a)

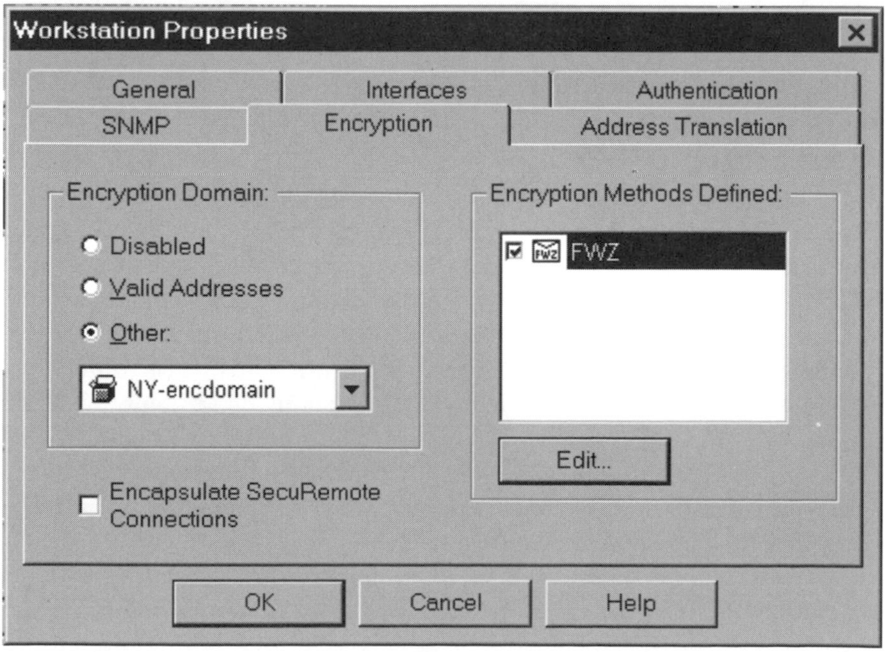

Workstation Properties

General	Interfaces	Authentication
SNMP	Encryption	Address Translation

Encryption Domain:
  ◯ Disabled
  ◯ Valid Addresses
  ◉ Other:

  📇 NY-encdomain ▼

  ☐ Encapsulate SecuRemote Connections

Encryption Methods Defined:
  ☑ 📇 FWZ

  [Edit...]

[OK]   [Cancel]   [Help]

(b)

Figure 7.49 Firewall-1 configuration at HQ for Internet VPN access.

3. Define remote access users using the "User" object and then configure the authentication scheme.

An example of the client encryption rule in the rule base is shown in Figure 7.50.

In the first rule, defined in the security policy shown in Figure 7.50, only the staff that belongs to the New York LAN has access to the services defined under the "service" column for the NY-encryption domain. The access is further restricted to office hours, as defined under the "Time" column. In this case, due to the action setting of "Client Encryption," all communications between the SecuRemote clients and the Firewall-1 server will be encrypted.

This completes the VPN setup process using the Checkpoint Firewall-1 based solution. It wasn't that hard, was it? Next we will use a Cisco router-based VPN solution for NIT's remote access needs.

Cisco Router-Based VPN Solution

Let's consider the example of a telecommuter or a branch office router setting a VPN over the public Internet using Cisco routers. We will refer to Figure 7.51 for this discussion.

Cisco provides various mechanisms to set up VPNs between two routers or between a router and a mobile client. These mechanisms are outlined below:

1. Using IPSec for setting up the encrypted tunnel;
2. Using Cisco proprietary Crypto tunnel solution.

Cisco broadly classifies its VPN solution based on who initiates the VPN tunnel. The tunnel may be client initiated or NAS initiated. The NAS-initiated tunnels are based on Cisco's implementation of layer two tunneling protocol (L2TP) or a Cisco proprietary Layer Two Forwarding (L2F), while the client-initiated VPN tunnels use the IPSec standard. For the purposes of this case study we will examine the configuration of a client-initiated IPSec tunnel.

In our case of a client-initiated tunnel, the tunnel end points are Cisco supported RavlinSoft Client on a laptop or a desktop and the Cisco router with IOS 11.3 at the New York site. The IOS 11.3 has built-in support for VPN and IPSec. The client in this case could also be a home or branch router with support for VPN. These two cases are shown in Figure 7.51.

Client-Initiated IPSec Tunnel

To set up a client-initiated IPSec tunnel, the user PC and the enterprise router need to support IPSec. For more details on IPSec, please refer to the discussion in Chapter 5. Cisco has partnered with RedCreek Communications for a

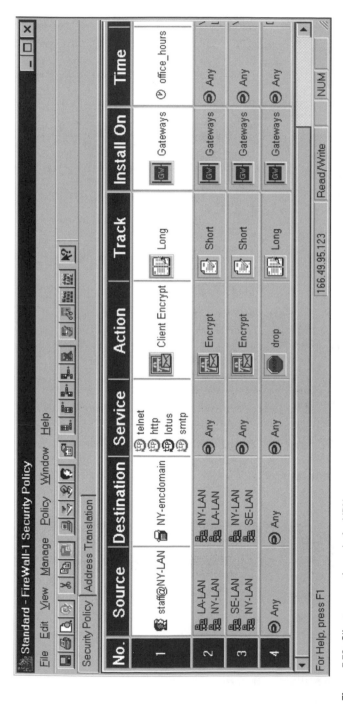

Figure 7.50 Client encryption rule for VPN access.

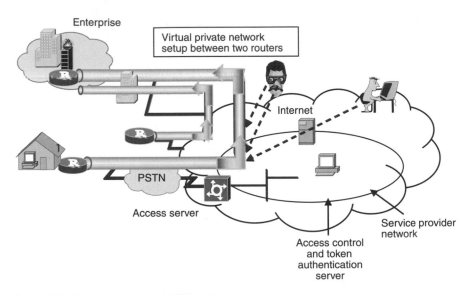

Enterprise

Virtual private network
setup between two routers

Internet

PSTN

Access server

Service provider
network

Access control
and token
authentication
server

Figure 7.51 Cisco router-based VPN architecture.

client-initiated IPSec solution. Cisco has incorporated RedCreek's software solution in Cisco IOS 11.3. For client software, RedCreek provides RavlinSoft
Remote Access Client software. Security administrators can configure various
VPN settings using Ravlin's own software manager. The various VPN settings
using IPSec are ESP tunnel mode, ESP transport mode, and RedCreek's own
"Encrypt-in-place" mode. The encryption is based on either 56-bit DES or
112/168-bit triple DES. The authentication is based on MD5 or SHA. An attractive component of authentication is support for X.509 v3-based certificate
authentication in the RedCreek solution. The certificates are signed using digital
signature standard (DSS) or SHA. The Cisco-RedCreek solution also adds support for DHCP and RADIUS-based user authentication. The natural advantage
of a router-based solution is that you do not have to install a separate firewall; the
obvious disadvantage is that this is not a scalable solution. The router-based solution will work only for a limited number of users. The number will depend on:

- The current load on the router;
- Router type and memory;
- Size of the routing table and ACL configuration;
- Length of the encryption key (56-bit or 168-bit);
- Whether the router performs authentication: if yes, then the length of
  the key is used;

- Size of SA database;
- Hashing algorithm used, MD5 or SHA.

For the best performance, use 56-bit DES, MD5 as the hashing algorithm, Diffie-Hellman algorithm for key exchange, and longer lifetime for the SA. The router platform will also play a key role. For example, Cisco 7513 router with 128 MB RAM and low CPU load can handle more IPSec tunnels than a Cisco 4000 router with 64 MB RAM and low CPU load. Due to a limited number of users today, NIT has chosen to test the remote access solution using the current Cisco 4700 serial router. We will now configure the router in the New York HQ to accept IPSec tunnel requests from dial-in end users.

The router configuration involves two steps. Step one is to perform the key exchange using IKE and step two is to configure the IPSec parameters like the tunnel mode, type of encryption, and method of authentication.

Configuration of IKE (Also Known as ISAKMP)

One way to configure IKE is to set up a manual key exchange, also called "presharing" of the keys. Alternatively, you can use a PKI to perform the key exchange. The PKI solution is much more scalable than the manual key exchange. You can set up your own CA and PKI infrastructure using software from Netscape, BBN, or Entrust. Alternatively, the CA service is available from VeriSign. We will use the CA-based configuration. For this example, we will assume that NIT has set up its own CA, known as NITCA. To use the CA method we need to follow the steps below:

- Create an RSA key pair for the router and the client.

```
1. R1# config terminal

2. R1(config)#

3. R1(config)# cry key gen rsa usage-keys

4. The name for the keys will be R1.nit.com

5. Choose the size of the key modulus in the range of
   360 to 2048 for your Signature key

6. How many bits in the modulus (512): ⏎

7. Generating RSA keys....

8. (OK)

9. Choose the size of the key modulus in the range of
   360 to 2048 for your Encryption key
```

```
10. How many bits in the modulus (512): ↵
11. Generating RSA keys
12. (OK)
13. R1(config)# ^z
14. R1# wri mem
```

The key setup parameter in this step is on line 3, "usage-keys." This command creates two RSA key pairs, one key pair for encryption and the other for digital signatures. If usage keys is not specified at the end, only one key pair is generated and is used for encryption and digital signatures.

- Request CA certificates.

In this step the router requests the CA certificates from the CA. If the router is configured with the domain and the DNS server information, you can ignore the following steps. Otherwise, it is a good idea to configure these addresses.

```
1. R1# config terminal
2. R1(config)#
3. R1(config)# ip host nitca 204.80.127.33
                              # Host name for NIT CA
4. R1(config)# ip domain-name nit.com
                              # Configuring the DNS
5. R1(config)# ip name-server 204.80.127.59
6. R1(config)# ip name-server 207.80.131.69
7. R1(config)# cry identity nit-ca
                              # Select a name for the key
8. R1(ca-identity)# enrollment url http ://nitca
                       # Use HTTP protocol and set URL
9. R1(ca-identity)# ex
10. R1(config)# crypto ca authenticate nit-ca
        # Instructs the router to get CA's  certificate
```

- Enroll certificates for the router with CA.

```
1. R1(config)# cry ca enroll nit-ca
```

```
                                        # Use HTTP protocol and set URL
 2. %start certificate enrollment ..

 3. %create a challenge password. You will need to ver-
    bally provide this password to the CA administra-
    tor in order to revoke your certificate. For secu-
    rity reasons your password will not be saved in the
    configuration. Please make a note of it.

 4. Password:

 5. Re-enter password:

 6. % The subject name for the keys will be: R1.nit.com

 7. %Include the router serial number in the subject
    name? (yes/no): yes

 8. %The serial number in the certificate will be :
    024070477

 9. %Include an IP address in the subject name?
    (yes/no): yes

10. Interface: Ethernet 0

11. Request certificate from CA? (yes/no): yes

12. R1(config)# ^z

13. R1# sh cry ca cert
```

At this stage, the command on line 13 above will show you both the CA and router certificate and their associated fingerprints. The CA certificate will be listed as "pending" at this time. Now contact the CA administrator by e-mail (PGP signed and encrypted) or by phone. Confirm the fingerprints of the CA certificate with the CA administrator. The administrator will confirm the fingerprints of the router's certificate that it has received for enrollment. Once the CA confirms the fingerprint and the host identity, it will issue a certificate. The status of the router certificate will change from "pending" to "available." This completes the CA enrollment process. Next we will configure the ISAKMP (IKE) policy objects.

- Configure IKE protection suite on NY router R1.

```
1. R1(config)# crypto isakmp policy 1

2. R1(config-isakmp)# hash md5

3. R1(config-isakmp)# lifetime 4000
```

```
4. R1(config-isakmp)# ex
5. R1(config)# ^Z
6. R1# wri mem
```

In this configuration, line 1 above configures one of the many ISAKMP (IKE) policy objects. In case of multiple policies, the policies are presented to the peer negotiating router in their numerical order. In this example we are using the default "rsa-sig" instead of "rsa-encr." The next step is to configure the IPSec parameters on the router.

Configuration of IPSec

Configuring IPSec involves following steps:

- Create IPSec transform set(s).

```
1. R1(config)# crypto ipsec transform-set remote1 ah-
   md5-hmac esp-des
2. R1(cfg-crypto-trans)# ex
3. R1(config)# crypto ipsec transform-set remote2 ah-
   rfc1828
4. R1(cfg-crypto-trans)# ex
5. R1(config)#
```

In this configuration we have configured two transform sets on lines 1 and 3 above. All transform sets will use the default tunnel mode. The transform set "remote1" uses both AH and ESP in the operation of an IPSec tunnel. The transform set "remote2" uses only the AH mode of operation. Both sets will be offered to the IPSec peer, but only one will be negotiated based on the preference of the peer.

- Create an extended ACL list for specifying the IPSec peers.

```
1. R1(config)# access-list 120 permit ip 207.80.131.0
   0.0.1.255 host 132.44.32.34
2. R1(config)# access-list 120 permit ip 132.44.32.0.0
   0.0.0.255 host 132.44.32.34
```

The list specifies all the addresses at remote locations of NIT. If NIT wants to permit access from anywhere on the Internet, then these access lists need to be modified as follows:

```
R1(config)#    access-list    120    permit    any    host
132.44.32.34
```

- Create dynamic crypto map.

```
1. R1(config)# crypto dynamic-map ny-router 10 ipsec-
   isakmp

2. R1(config-crypto-map)#   set   security-association
   lifetime seconds 4000

3. R1(config-crypto-map)#   set   transform-set   remote1
   remote2

4. R1(config-crypto-map)# match address 120

5. R1(config-crypto-map)# ex

6. R1(config)#
```

Dynamic crypto maps allow the router to set up dynamic associations with unknown peers. In case of mobile clients logging onto the router, this would be the most appropriate configuration. Dynamic crypto maps allow routers to fill-in information based on dynamic negotiation of the IPSec security association parameters. Once the lifetime of a security association expires (configured for 4,000 seconds in line 2 above), the temporary crypto map entries are removed. The transform-sets are offered in the order listed in line 3 above. The inbound traffic is then matched against the extended access list 120. Add the following ACL to prevent the subnet broadcast traffic from being dropped.

R1(config)# access-list 120 deny ip 207.80.131.0 0.0.1.255 132.44.32.255 0.0.0.0

- Apply crypto map to appropriate interface (in this example the e1 interface).

```
1. R1(config)# int e 1

2. R1(config-if)# crypto map ny-router

3. R1(config-if)#ex

4. R1(config)# ^Z

5. R1# wri mem
```

Following is the final configuration of the router.

```
R1#wri t
Building configuration...
Current configuration:
!
! Last configuration change at 10:45:17 EDT Fri Oct 16 1998
by xxxx
! NVRAM config last updated at 17:20:20 EDT Mon Jun 15 1998
by xxxx
!
version 11.3
no service finger
no service pad
service timestamps debug datetime msec localtime show-
timezone
service timestamps log datetime msec localtime show-time-
zone
service password-encryption
service compress-config
no service udp-small-servers
no service tcp-small-servers
!
hostname R1
!
clock timezone EDT -5
clock summer-time EDT recurring
boot system flash
boot system rom
enable secret 5 yyyy
enable password 7 ye4rftz3e2w
!
ip host nitca 204.80.127.33          # Host name for NIT CA
ip domain-name nit.com               # DNS server entries
ip name-server 204.80.127.59
ip name-server 207.80.131.69
!
crypto isakmp policy 1               # ISAKMP policy objects
   hash md5
   lifetime 4000
!
crypto dynamic-map ny-router 10 ipsec-isakmp
   set security-association lifetime seconds 4000
```

```
   set transform-set remote1 remote2
   match address 120
  !
ip subnet-zero
no ip source-route
!
.
.(Configuration listed in case study five)
.
!
interface Ethernet0   ## This sets the ethernet interface
  description Connection to the admin subnet
  ip address 207.80.127.36 255.255.255.224
          ## Sets the interface IP  address with the mask
  ip broadcast-address 207.80.127.63
            ## The mask is a VLSM  creating 32 host subnet
  ip access-group 131 out ## Sets the ACL on this interface
  no ip redirects
  no ip proxy-arp
  no keepalive
!
interface Ethernet1
description Sales and Customer service subnet
  ip address 207.80.127.71 255.255.255.224
          ## Sets the interface IP  address with the mask.
  ip broadcast-address 207.80.127.95
            ## The mask is a VLSM  creating 32 host subnet.
  ip access-group 132 out ## Sets the ACL on this interface
  crypto map ny-router
no ip redirects
  no ip proxy-arp
  no keepalive
!
access-list 120 permit ip 207.80.131.0 0.0.1.255 host
132.44.32.34
access-list 120 permit ip 132.44.32.0.0 0.0.0.255 host
132.44.32.34
access-list  120  deny  ip  207.80.131.0  0.0.1.255
132.44.32.255 0.0.0.0
!
access-list 131 permit tcp any any established
```

```
                      ## Passes all the established connections
access-list   131   deny   ip   207.80.127.32   0.0.0.31
207.80.127.32 0.0.0.31              ## Antispoofing Filter
## This is also the first time you will observe the in-
verse masking technique used by  Cisco.
access-list 131 permit tcp 207.80.128.0 0.0.0.255 any eq
www  ## Allows all the web connections on Port 80 from
    Chicago LAN
access-list 131 permit tcp 207.80.129.0 0.0.0.255 any eq
www ## Allows all the  web connections on Port 80 from
    Boston LAN
access-list 131 permit tcp 207.80.130.0 0.0.0.255 any eq
www  ## Allows all the web connections on Port 80 from DC
    LAN
access-list 131 permit tcp 207.80.128.0 0.0.0.255 any eq
443  ## Allows all the secure web connections on Port  443
    from Chicago LAN
access-list 131 permit tcp 207.80.129.0 0.0.0.255 any eq
443 ## Allows all the web connections on Port 80 from
    Boston LAN
access-list 131 permit tcp 207.80.130.0 0.0.0.255 any eq
443 ## Allows all the web connections on Port 80 from DC
    LAN
access-list 131 permit tcp any any eq 53
      ## Allows all the DNS  connections on TCP Port 53
access-list 131 permit udp any any eq 53
        ## Allows all the DNS connections on UDP Port 53
access-list 131 permit tcp any any eq 25
          ## Allows all the SMTP connections on Port 25
access-list 131 permit tcp any any eq 110
         ## Allows all the POP3 connections on Port 110
access-list 131 permit icmp any any
        ## Allows all the ICMP connections to the subnet
access-list 132 permit tcp any any established
        ## Passes all the established connections
access-list   132   deny   ip   204.80.127.64   0.0.0.31
204.80.127.64 0.0.0.31              ## Antispoofing Filter
access-list 132 permit tcp 204.80.129.0 0.0.0.255 any eq
telnet##Allows all the telnet  connections from Boston LAN
access-list 132 deny icmp any any
!
```

```
 snmp-server  community  goodsnmp
!
line  con  0
 exec-timeout  15  0
 password  7  yzcv45rfs6fgt$#eth%
!
line  vty  0
 access-class  109  in
 access-class  109  out
 exec-timeout  15  0
 password  7  6dtysb%3s8$frdtsr
!
 line  vty  1  4
exec-timeout  15  0
 password  7  5thrs^srfv78fvsdt5t
 !
end
```

This completes the configuration of router R1 for accepting a IPSec tunnel request from a mobile user. The security administrators at NIT now need to install the RavlinSoft Client on the remote desktops and register their keys with the NIT CA. Once this step is completed, the remote users can set up IPSec VPN tunnels with the NIT router. The same principle applies to remote branch routers. The remote routers will need to have Cisco IOS 11.3 and will have to follow the steps used to configure router R1 in this case study.

7.6.3 Conclusion

We started the case study with the objective of setting VPN tunnels using IPSec technology. We configured the IPSec tunnels using a Cisco and Checkpoint implementation of IPSec. Since IPSec is a standard track framework, all the IPSec implementations are supposed to be interoperable. Test the vendor software for IPSec interoperability if you plan to have a mixed environment of IPSec implementations.

7.7 Case Study VII: Extranet Access via VPN

7.7.1 Case Study Objective

At the end of this case study you will be able to set up a secure extranet connection for a vendor or a partner over the Internet using a combination of

SOCKSv5 and SSL3.0. In this case study we will use the extranet solution from Aventail. We will explore the client- and server-side configuration for the Aventail extranet solution.

7.7.2   Case Study Background and Requirements

NIT would like to set up a secure extranet connection with its partners in the United States. The extranet setup has to be simple to set up, easy to administer, and provide a secure VPN connection over the public Internet. The initial goal of the extranet setup is to provide partner access to NIT's Web-based ticketing and customer information system on the intranet. The extranet Web server communicates with the backend database server for on-line transaction. Please refer to the architecture section for further details. NIT would like to keep a log of partner access for audit purposes.

Architecture

There are various ways to set up an extranet connection between a host and a partner over the Internet as we have seen in Chapter 5 on security and VPN. In the previous case study and in the chapter on security, the two fundamental requirements for setting up a VPN connection are data encryption and end-point authentication. In the previous case study we based our solution on the IPSec standard, while in this case study we will use another standards-based solution which uses IETF standardized SOCKSv5 and industry standardized SSL protocol.

The Aventail solution uses both SOCKS5 and SSL standards to create a VPN solution that can be used to set up an authenticated and secure extranet connection between two partners as required in our case study. The architecture for such a setup is shown in Figure 7.52.

All external connections are channeled through Aventail's Extranet Center Firewall. Each NIT partner sets up an authenticated SSL tunnel between its firewall, or a user workstation, to the NIT firewall. The authentication could be based on X.509-based certificates or a simple userID/password mechanism. The traditional userID/password mechanism is used as a backup authentication mechanism. In this architecture we are assuming a LAN-to-LAN setup, though the solution can also accommodate a dial-up extranet connection.

The Aventail extranet solution consists of two main components:

1. A SOCKSv5-based proxy server;
2. A client software for Windows platform.

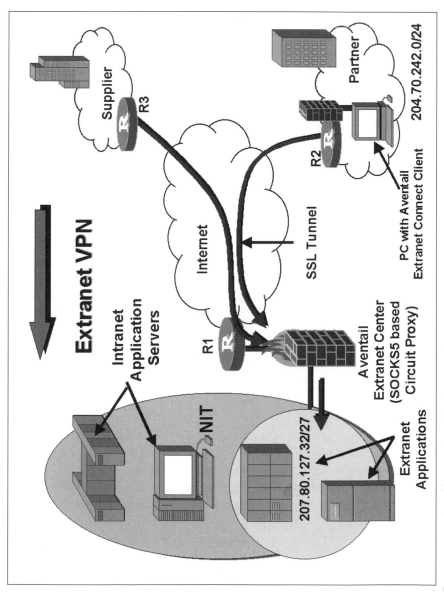

Figure 7.52 Extranet network architecture based on Internet VPN.

Additional components are the policy console for administering the security policy and the management server for remotely administering the firewall.

SOCKSv5-Based Proxy Server

As we have seen in the chapter on security, SOCKSv5 provides a "circuit-proxy" at the session layer of an OSI stack. Thus, unlike application proxy firewalls, the Aventail SOCKSv5-based solution is application independent and offers more granular access control features than at the IP layer. Also, unlike the previous incarnation of SOCKS, version 4, SOCKS version 5 supports:

- Client authentication based on userID/password, CHAP, S/Key, digital certificates, CRAM, and token cards (e.g., security dynamics);
- Traffic encryption (with SSL);
- Support for UDP traffic.

The Aventail firewall based on the SOCKSv5 standard offers a proxy service for all internal applications and channels all the incoming requests through TCP port 1080. All incoming connections are filtered and authenticated based on the security policy implemented through a policy console. For example, in Figure 7.53, the first rule in the "Policy Console GUI" states that all incoming connections with destination of "Extranet_Web" server are allowed to pass through if they pass the SSL authentication test, otherwise the connections are dropped.

In Figure 7.53, the inside window titled "access control builder" shows various destination networks available for constructing the access control rule. The "selected services" window shows that the port 80 has been selected for the "Extranet_Web" server.

Since all connections are channeled through the SOCKSv5 proxy, it could introduce performance issues as well as a single point-of-failure. To avoid this scenario, the Aventail Extranet Center can be configured in a cluster configuration with a shared file system and a third party load-balancing solution from BIG/ip, Alteon, or Cisco Local Director. This architecture is shown in Figure 7.54.

The Client Software

It is the client software on user desktop that directs an application request received via Winsock to the appropriate SOCKSv5 proxy server. The client software runs transparently on each desktop and sits between Winsock and the TCP/IP layer (see Figure 7.55).

An Aventail client can modify data as it passes from application layer to the TCP/IP layer. An example could be data encryption or data compression.

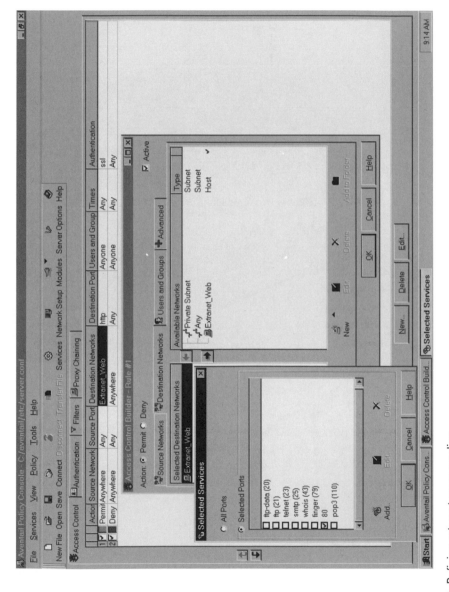

Figure 7.53 Defining extranet access policy.

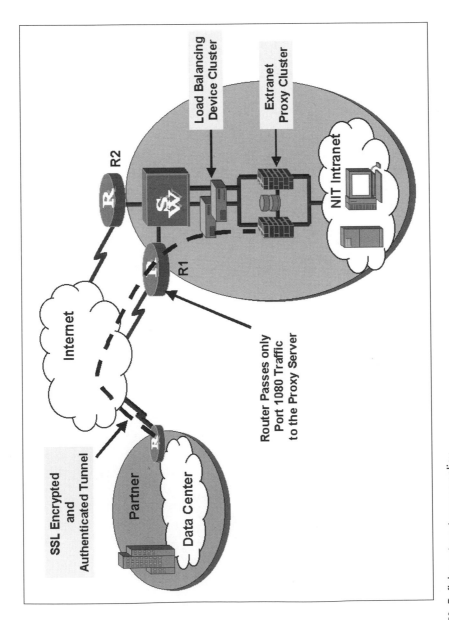

Figure 7.54 Defining extranet access policy.

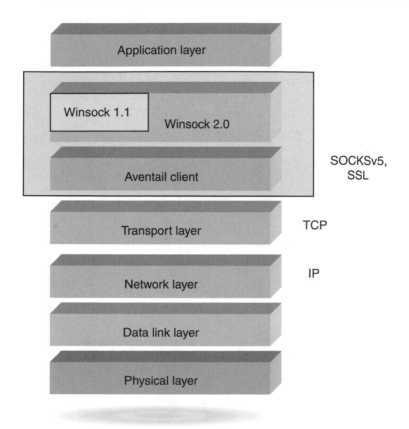

Figure 7.55 Proxy operation mapped to OSI model.

It is important to note that the Aventail client does not replace the Windows Winsock nor does it replace or modify the TCP/IP stack in any way. The Aventail client configuration tool provides a GUI-based tool for configuration of the client (see Figure 7.56).

The first window in Figure 7.56 shows the "Redirection Rule" at the client end that redirects all traffic destined to the "Private Network" to the "Aventail Extranet" server. All other traffic is passed as normal traffic without any redirection. The second screen shows the "Authentication" configuration GUI with options for various client authentication schemes including CHAP, CRAM, SSL3.0, SOCKSv4 Auth, and HTTP basic authentication based on RFC 2068.

In the next section we will implement the NIT extranet architecture shown in Figure 7.52 using the Aventail Extranet Proxy Server and the Aventail Connect Client.

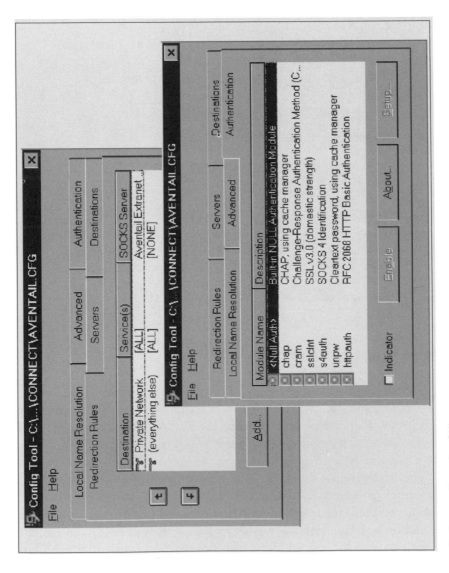

Figure 7.56 Aventail client configuration GUI.

Implementation Details of NIT Extranet

The NIT extranet implementation is a two-stage process. In the first stage, the Aventail Extranet Proxy Server is installed and configured on a UNIX workstation at the NIT site. The ACL on the NIT router is then configured for allowed port 1080 access from extranet partners.

In the second stage, the Aventail Extranet Connect Client is installed and configured on the partner Windows 98 workstation.

The server hardware and software configuration for Extranet Proxy Firewall is as follows.

Sun Ultra Sparc 2 server with:

- 256 MB RAM;
- 8 GB hard drive;
- 19-inch monitor;
- 10/100 MB ethernet interface;
- QTY 2 (one interface is standard on the system board);
- 4-mm tape drive;
- QTY 1;
- Internal CD-ROM drive;
- Sun Solaris 2.5.1 or higher.

Software: Aventail Extranet Center version 3.0 and Aventail Extranet Connect version 3.0

Aventail Proxy Server Installation Process

Step 1   Install the Aventail Extranet Center CD in the CD-ROM drive of the Sun Proxy Server.

Step 2   Proxy# cd/aventail/server

Step 3   Proxy# ./install.sh   *!Run the installation script provided by Aventail*

Detecting operating system/architecture ... apparently solaris/sparc
Installing to prefix: /usr/local/aventail   *!Default installation directory. If you want to change this default setting use a switch with the installation script Ex. Install.sh −prefix => /install directory>/*

AVENTAIL EXTRANET CENTER   *!License agreement*
END USER LICENSE AGREEMENT

IMPORTANT - READ CAREFULLY

...

....

....

Step 4   Do you agree to this license agreement (y/n) y

Step 5   Would you like to install Aventail S5 Server v3.00? y
          Installing package server in /usr/local/aventail... done.

Step 6   Would you like to install Various SOCKSified applications? y
          Installing package clients in /usr/local/aventail... done.

Step 7   Would you like to install Administration Tool? y
          Installing package config in /usr/local/aventail... done.

Step 8   Would you like to install Online help for administration tool? y
          Installing package help in /usr/local/aventail... done.

Step 9   Would you like to install Management Server? y
          Installing package mgmtserv in /usr/local/aventail... done.
          Installing package modules in /usr/local/aventail... done.
          Installing package messages in /usr/local/aventail... done.
          Installing package tools in /usr/local/aventail... done.
          Installing package manpages in /usr/local/aventail... done.

Step 10   Would you like to install Sample configuration files (server)? y
          Installing package sample in /usr/local/aventail... done.

Step 11   Installing package qt in /usr/local/aventail... done.

Step 12   In order for the GUI tools to work, please
          set the **LD_LIBRARY_PATH** variable to include /usr/local/
          aventail/lib.
          csh/tcsh: setenv LD_LIBRARY_PATH/usr/local/aventail/
          lib:$LD_LIBRARY_PATH
          sh/bash: LD_LIBRARY_PATH=/usr/local/aventail/lib:
          $LD_LIBRARY_PATH ; export LD_LIBRARY_PATH

Step 13   A license file for server should have been e-mailed to you (or accompanied a CD-ROM distribution). Please install this license file in /usr/local/aventail/etc/aventail.alf.

Installation complete.

Step 14   You are now ready to configure your server.

Step 15   # cp   /apps/kphaltan/aventail1.alf   /usr/local/aventail/etc/
           !Copy the license file to appropriate directory

Step 16   # cd /usr/local/aventail/etc/

Step 17   # ls

           about.xpm   **aventail1.alf** s5.conf   splash16.xpm

           about16.xpm   manager.conf   splash.xpm

Step 18   Please run:

           /usr/local/aventail/bin/apc

           under X-Windows to start the console.

Step 19   The Policy Console will show the display shown in Figure 7.57.

The first policy rule that is installed by default is the "deny all" rule.

Step 20   In order to define the security policy via the policy console, the next step is to define various network objects of interest. For example:

- Network objects like the NIT intranet LAN segment or the partner LAN segment;
- Objects like the extranet Web server or a specific client desktop identified by its fully qualified domain name (FQDN) or IP address. Figure 7.58 shows the policy console screen and the menu options with a policy setup.

You can define host, domain, subnet or a range of IP addresses as the policy objects. Once defined, these policy objects can be used to set up granular access policies. An example is seen in Figure 7.58, where access to "Extranet_Web" object is further restricted to services on port 80 only.

Step 21   We are now ready to configure the router R1 at the NIT location with appropriate access control lists. The access control rules on router R1 need to permit all access from the partner network (204.70.242.0/24) to the Aventail Extranet Proxy Server (207.80.127.39/27) on port 1080. The access list on R1 would look like:

Figure 7.57   Aventail policy GUI.

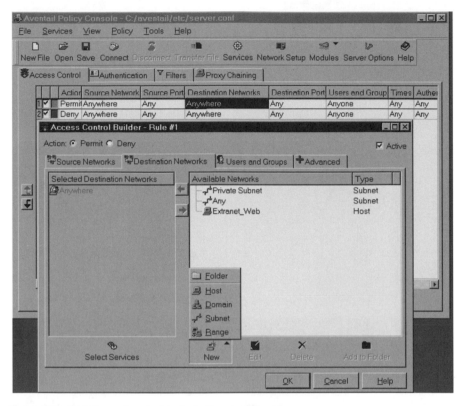

Figure 7.58 Defining network objects in the security policy.

> *access-list 136 deny ip 207.80.127.0 0.0.0.255 204.80.127.0 0.0.0.255*
>
> *access-list 136 permit tcp 204.70.242.0 0.0.0.255 host 207.80.127.39 eq 1080*
>
> *access-list 136 deny ip 204.70.242.0 0.0.0.255 any*

The first access control rule is an antispoofing filter, while the second rule filters the traffic for port 1080 from the partner network.

We are now ready to install and configure the Aventail Extranet Connect Client on the partner network.

Aventail Connect Client Installation

This is a basic installation process for the client software. For advanced configuration, please refer to the Aventail installation guide.

Step 1   Under Windows, install the Aventail connect software by running the "set up.exe" file.

Step 2   Install the appropriate license file by copying "aventail.alf" file to the home directory of installation.

Step 3   Configure the client software by running the Configuration Tool. The first configuration step is to define the extranet SOCKS proxy server (see Figure 7.59).

Step 4   The next step is to define the destinations and the redirection rule for all traffic destined from the client workstation to the destination network. In this case study, the destination network is the NIT intranet and the redirection rule on the Aventail client needs to indicate the NIT proxy server as the target proxy server to reach the NIT network (see Figure 7.60). For all other traffic not destined to the NIT intranet, the traffic is routed as normal, without redirection.

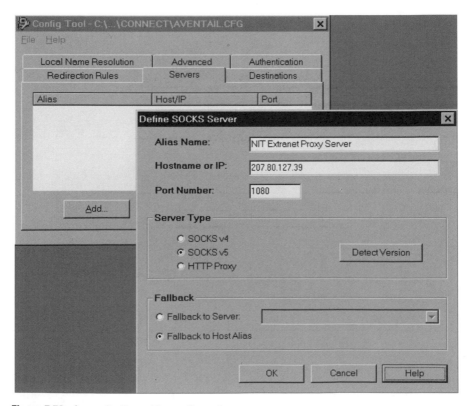

Figure 7.59 Aventail client-side configuration.

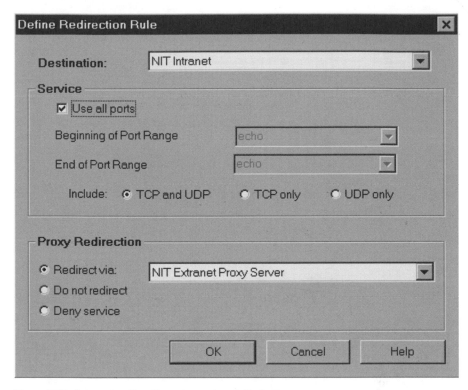

Figure 7.60 Extranet redirection rule on Aventail client.

Step 5 Restart the Aventail client software with the new configuration in-formation.

Step 6 When a user on the partner network tries to connect to the NIT ex-tranet Web server, traffic exits the partner network in one of three ways:

1. Unrestricted outbound connection on partner's firewall.

2. Partner uses its own Aventail SOCKS server to pass traffic to the NIT SOCKS server. This set up SOCKS to SOCKS communi-cation.

3. Traffic exits partner network via a local HTTP proxy server (SOCKS over HTTP).

Step 7 When a user on the partner network connects to the NIT extranet web server, it is presented with the extranet server certificate (see Figure 7.61). For example, the certificate in Figure 7.61 has been issued to NIT by an independent CA called NetPlexus. The certificate is a "server" certificate for server 207.80.127.39.

Step 8   If the certificate is acceptable, the user clicks on Accept.

Step 9   The client and server create an SSL connection as described in Chapter 5.

Step 10   The user is then presented with an appropriate authentication screen by the server. The authentication could be based on userID/password, user digital certificate, or a token-based password (see Figure 7.61).

Step 11   If the authentication is successful, the authentication credentials can be optionally "cached" for the duration of the session on the user workstation, thus avoiding the need to perform frequent authentication for every new session.

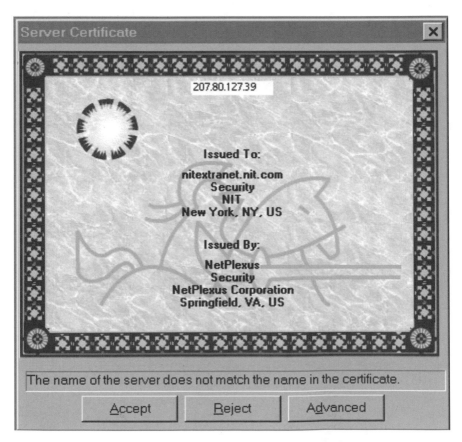

Figure 7.61 Server certificate for secure extranet connection using SSL.

Step 12   The extranet authenticated SSL connection is now functional between the partner network and the NIT extranet server.

7.7.3 Conclusion

This completes our basic configuration of an NIT extranet setup and the series of case studies used to illustrate the configuration details of implementing secure intranets and extranets.

Throughout this case study we have looked at the implementation of network components like routers, switches, dial-access servers, and security components such as the access control rules on the routers and configuration of virtual private networks using IPSec compliant firewalls and routers. In the end, we have concluded with a SOCKSv5-based proxy firewall and SSL-based extranet implementation.

List of Acronyms and Abbreviations

AAL	ATM adaption layer
ACFC	address-and-control field compression
ACL	access control list
ADSL	asymmetric digital subscriber line
AH	authentication header
AMI	alternate mark inversion
ANX	automotive network exchange
ARP	address resolution protocol
AS	autonomous systems
ASIC	application specific integrated circuit
ASN	abstract syntax notation
ATM	asynchronous transfer mode
AUI	auxiliary user interface
AUP	acceptable use policy
AuthS	authentication server
BGP	border gateway protocol
BRI	basic rate interface
BUS	broadcast and unknown server
CA	certifying authority
CAD	computer aided design
CBL	common business lines,
CBR	constant bit rate
CDDL	copper distributed data interface
CERT	Computer Emergency Response Team
CGI	common gateway interface
CHAP	challenge handshake authentication protocol

CIM	common information model
CIP	channel interface processor
CIR	committed information rate
CLEC	competitive local exchange carrier
CLP	cell loss priority
CMIP	common management information protocol
CPE	customer premise equipment
CPU	central processing unit
CRAM	challenge-response authentication method
CRC	cyclic redundancy check
CS	convergence sublayer
CSMA/CD	carrier sense multiple access and collision detection
CSU/DSU	channel service unit/data service unit
CUG	closed user groups
DAC	dual attached concentrator
DAS	dual attached station
DCE	distributed computing environment
DE	discard eligible
DES	data encryption standard
DHCP	dynamic host configuration protocol
DLCI	data link connection identifier
DLSw	data link switching
DMTF	desktop management force
DMZ	de-militarized zone
DNIS	dialed number information
DoS	denial of service
DQDB	dual queue dual bus
DSS	digital signature standard
DUAL	diffusing update algorithm
DXI	data exchange interface
ECA	ESCON channel adapter
EDI	electronic data interchange
EIGRP	enhanced IGRP
ELAN	emulated LAN
ELP	ethernet interface processor
ERP	enterprise resource planning
ESCON	enterprise systems connection
ESF	extended super frame
ESP	encapsulating security payload
FCIRT	federal computer incident response team
FCS	frame check sequence

FDDI	fiber distributed data interface
FEP	front-end processor
FQDN	fully qualified domain name
FRAD	frame relay access device
FSIP	fast serial interface processor
FTP	file transfer protocol
GRE	generic routing encapsulation
GSSAPI	generic security service application program interface
GUI	graphical user interface
HDLC	high level data link control
HSRP	hot-standby routing protocol
HSSI	high-speed serial interface
HTML	hypertext markup language
HTTP	hypertext transport protocol
IAB	Internet Architecture Board
IANA	Internet Assigned Numbers Authority
ICMP	Internet control message protocol
IEEE	institute of electrical and electronic engineers
IETF	Internet Engineering Task Force
IGP	interior gateway protocol
IGRP	interior gateway routing protocol
IKE	Internet key exchange
ILMI	interim local management interface
IPSec	IP security
IPX	internetwork packet exchange
IS-IS	intermediate system-to-intermediate system
ISAKMP	Internet Security Association and Key Management Protocol
ISDN	integrated services digital network
ISL	inter switch link
ISP	Internet service provider
IT	information technology
KDC	key distribution center
L2TP	layer two tunneling protocol
LAN	local area network
LAC	layer 2 access concentrator
LANE	LAN emulation
LAP-B	link access procedure balanced
LAT	local access transport
LCP	link control phase
LDAP	lightweight directory access protocol
LEC	LAN emulation client

LEC	local exchange carrier
LECS	LAN emulation configuration server
LES	LAN emulation server
LLC	logical link control
LNS	layer 2 network server
LSA	link state advertisement
LSP	layered service provider
LUNI	LEC user-to-network interface
MAC	media access control
MAN	metropolitan area network
Mbps	mega bits per second
MD5	message digest 5
MII	media independent interface
MIME	multi-purpose Internet mail extension
MIS	management information base
MPLS	multiprotocol label switching
MPOA	Multiprotocol over ATM
MTU	maximum transmission unit
MUX	multiplexer
NAP	network access point
NAS	network access server
NAT	network address translation
NCP	network control phase
NIC	network interface card
NMS	network management system
NNI	network-to-network interface
NNM	network node manager
NOC	Network Operations Center
NSAP	network service access points
NTU	network terminating equipment
OC	optical connect
OID	object ID
OSI	open systems interconnection
OSPF	open shortest path first
OTP	one-time passwords
PAP	password authentication protocol
PCA	parallel channel adapter
PCI	protocol control information
PDU	protocol data unit
PFC	protocol field compression
PKI	public key infrastructure

PLP	packet layer protocol
PNNI	private network-to-network interface
POP	point-of-presence
POP3	post office protocol version 3
PPP	point-to-point protocol
PPS	packets-per-second
PPTP	point-to-point tunneling protocol
PRI	primary rate interface
PSTN	public switched telephone network
PVC	permanent virtual circuits
PVC	private virtual circuit
QOS	quality of service
RADIUS	Remote Authentication Dial In User Service
RBOC	regional bell operating company
RDBMS	relational database management system
RFI	request for information
RFP	request for proposal
RIP	routing information protocol
RMON MIB	remote network monitoring MIB
ROI	return-on-investment
RSVP	resource reservation protocol
RTMP	routing table maintenance protocol
SA	security association
SAR	segmentation and reassembly
SAS	single attached station
SDH	synchronous digital hierarchy
SDLC	synchronous data link control
SF	super frame
SHA	secure hash algorithm
SIP	SMDS interface protocol
SKIP	simple key management for IP
SLA	service level agreement
SLIP	serial link internet protocol
SMDS	switched multimegabit data service
SMI	structured management information
SMTP	simple mail transport protocol
SNA	systems network architecture
SNI	service-to-network interface
SNMP	simple network management protocol
SOHO	small office home office
SONET	synchronous optical network

SPI	security parameter index
SPID	service profile ID
SPX	sequenced packet exchange
SRB	source route bridging
SSL	secure socket layer
SSO	single sign-on
SS7	signaling system number seven
STP	shielded twisted pair
SVC	switched virtual circuit
TA	terminal adapter
TACACS	terminal access controller access
TCP/IP	transmission control protocol/Internet protocol
TDM	time division multiplexing
Telnet	for remote login
TGS	ticket-granting server
THT	token holding time
TTL	time to live
TTRT	target token rotation timer
UDP	user datagram protocol
UNI	user-to-network-interface
UPS	uninterrupted power supply
URL	uniform resource locator
UTP	unshielded twisted pair
VLAN	virtual LAN
VBR	variable bit rate
VCC	dynamic virtual channel circuit
VC	virtual circuit
VLSM	variable length subnet mask
VP	virtual path
VPDN	virtual private dial network
VPI	virtual path identifier
VPN	virtual private networking
VTP	virtual trunk protocol
XML	extended markup language

Bibliography

Breyer, Robert, and Sean Riley. *Switched, Fast, and Gigabit Ethernet*, 3d ed. Macmillan Technical Publishing, 1999.

Cavanagh, P. James. *Frame Relay Applications: Business and Technology Case Studies*. San Francisco: Morgan Kaufmann Publishers, 1998.

Comer, D. *Internetworking with TCP/IP: Principles, Protocols, and Architectures*. Englewood Cliffs, NJ: Prentice-Hall, 1988.

Feit, Sidnie. *SNMP: A Guide to Network Management*. New York: McGraw-Hill, 1995.

Garfinkel, Simson, and Gene Spafford. *Web Security and Commerce*. Cambridge, MA: O'Reilly, 1997.

Ginsburg, David. *ATM: Solutions for Enterprise Internetworking*. Reading, MA: Addison Wesley, 1996.

Halabi, Bassam. *Internet Routing Architectures*. Indianapolis, IN: Cisco Press and New Riders, 1997.

Huitema, Christian. *Routing in the Internet*. Englewood Cliffs, NJ: Prentice-Hall, 1995

Johnson, W. Howard. *Fast Ethernet: Dawn of a New Network*. Englewood Cliffs, NJ: Prentice-Hall, 1996.

Kercheval, Berry. *TCP/IP over ATM: a No-Nonsense Internetworking Guide*. Englewood Cliffs, NJ: Prentice-Hall, 1998.

Klessing, Tesink. *SMDS*. Englewood Cliffs, NJ: Prentice-Hall, 1995.

Mirchandani Sonu, and Khanna Raman. *FDDI: Technology and Applications*. New York: Wiley, 1993.

Perlman, Radia. *Interconnections: Bridges and Routers*. Reading, MA: Addison Wesley, 1992.

Ranade, Jay, and Sackett, George C. *Introduction to SNA Networking*, 2nd ed. New York: McGraw-Hill, 1995.

Rose, M. T. *The Simple Book: An Introduction to Management of TCP/IP-based Internets*. Englewood Cliffs, NJ: Prentice-Hall, 1991.

Schneier, Bruce. *Applied Cryptography: Protocols, Algorithms, and Source Code in C*. New York: Wiley, 1994.

Scott, Charlie, Paul Wolfe, Mike Erin. *Virtual Private Networks*. Cambridge, MA: O'Reilly, 1998.

Smith, Philip. *Frame Relay: Principles and Applications*. Reading, MA: Addison Wesley, 1996.

Solomon, D. James. *Mobile IP: The Internet Unplugged*. Englewood Cliffs, NJ: Prentice-Hall, 1998.

Spohn, L. Darren. *Data Network Design*, 2nd ed. New York: McGraw-Hill, 1997.

Stallings, William. *Network and Internetwork Security: Principles and Practice*. Englewood Cliffs, NJ: Prentice-Hall, 1995.

Stevens, W. Richard. *TCP/IP Illustrated: Volume 1–3*. Reading, MA: Addison Wesley, 1994.

About the Author

Kaustubh Phaltankar has extensive experience in the Internet industry. His new venture, www.netplexus.com, provides managed VPN and firewall services to enterprises who wish to outsource the management and deployment of enterprise intranet and extranets. Kaustubh also manages an information resource site, www.vpnmarket.com, for IT professionals planning to use VPN in their intranet and extranet implementations.

Prior to founding NetPlexus, Kaustubh held the position of Chief Engineer and Chief Architect at MCI's Internet Solutions Center in Virginia. Kaustubh was instrumental in designing the architecture for MCI's Web hosting center and has patents pending in this area. He has designed and implemented numerous Internet, intranet, and extranet solutions for MCI's Fortune 100 customers. He has worked on national and international projects with British Telecom, Concert Communications, and Bell Canada.

Kaustubh holds an M.S. degree in telecommunications and computer science from George Washington University, Washington D.C. He has also taught in the undergraduate electrical engineering program at the University.

Index

Super-High-Definition Images: Beyond HDTV, Naohisa Ohta

Telecommunications Department Management, Robert A. Gable

Telecommunications Deregulation, James Shaw

Telemetry Systems Design, Frank Carden

Teletraffic Technologies in ATM Networks, Hiroshi Saito

Understanding Modern Telecommunications and the Information Superhighway, John G. Nellist and Elliott M. Gilbert

Understanding Networking Technology: Concepts, Terms, and Trends, Second Edition, Mark Norris

Understanding Token Ring: Protocols and Standards, James T. Carlo, Robert D. Love, Michael S. Siegel, and Kenneth T. Wilson

Videoconferencing and Videotelephony: Technology and Standards, Second Edition, Richard Schaphorst

Visual Telephony, Edward A. Daly and Kathleen J. Hansell

Wide-Area Data Network Performance Engineering, Robert G. Cole and Ravi Ramaswamy

Winning Telco Customers Using Marketing Databases, Rob Mattison

World-Class Telecommunications Service Development, Ellen P. Ward

For further information on these and other Artech House titles, including previously considered out-of-print books now available through our In-Print-Forever® (IPF®) program, contact:

Artech House	Artech House
685 Canton Street	46 Gillingham Street
Norwood, MA 02062	London SW1V 1AH UK
Phone: 781-769-9750	Phone: +44 (0)20 7596-8750
Fax: 781-769-6334	Fax: +44 (0)20 7630-0166
e-mail: artech@artechhouse.com	e-mail: artech-uk@artechhouse.com

Find us on the World Wide Web at:
www.artechhouse.com